OCR gateway

GCSE biology

Authors

Simon Broadley

Sue Hocking

Mark Matthews

Contents

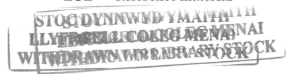

How to use this book

Welcome to your Gateway Biology course. This book has been specially written by experienced teachers and examiners to match the 2011 specification.

On these two pages you can see the types of pages you will find in this book, and the features on them. Everything in the book is designed to provide you with the support you need to help you prepare for your examinations and achieve your best.

Module openers

Specification matching grid: This shows you how the pages in the module match to the exam specification for GCSE Biology, so you can track your progress through the module as you learn.

Why study this module: Here you can read about the reasons why the science you're about to learn is relevant to your everyday life.

You should remember: This list is a summary of the things you've already learnt that will come up again in this module. Check through them in advance and see if there is anything that you need to recap on before you get started.

Opener image: Every module starts with a picture and information on a new or interesting piece of science that relates to what you're about to learn.

Main pages

Learning objectives: You can use these objectives to understand what you need to learn to prepare for your exams. Higher Tier only objectives appear in pink text.

Key words: These are the terms you need to understand for your exams. You can look for these words in the text in bold or check the glossary to see what they mean.

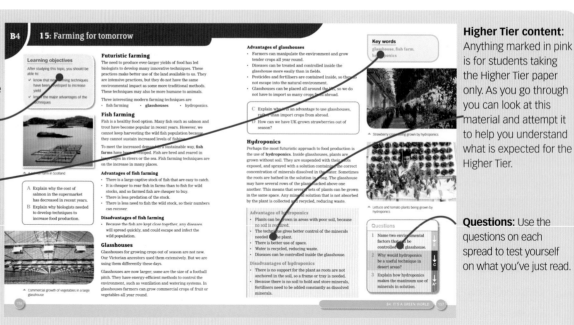

Higher Tier content: Anything marked in pink is for students taking the Higher Tier paper only. As you go through you can look at this material and attempt it to help you understand what is expected for the Higher Tier.

Questions: Use the questions on each spread to test yourself on what you've just read.

Summary and exam-style questions

Every summary question at the end of a spread includes an indication of how hard it is. You can track your own progress by seeing which of the questions you can answer easily, and which you have difficulty with.

When you reach the end of a module you can use the exam-style questions to test how well you know what you've just learnt. Each question has a grade band next to it, so you can see what you need to do for the grade you are aiming for.

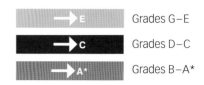

→ E	Grades G–E
→ C	Grades D–C
→ A*	Grades B–A*

Revision checklist:

This is a summary of the main ideas in the module. You can use it as a starting point for revision, to check that you know about the big ideas covered.

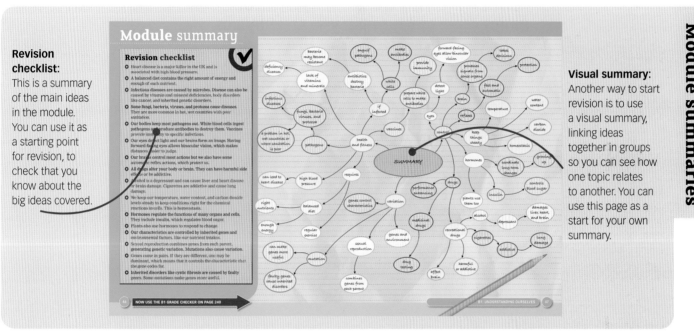

Visual summary:

Another way to start revision is to use a visual summary, linking ideas together in groups so you can see how one topic relates to another. You can use this page as a start for your own summary.

Upgrade:

Upgrade takes you through an exam question in a step-by-step way, showing you why different answers get different grades. Using the tips on the page you can make sure you achieve your best by understanding what each question needs.

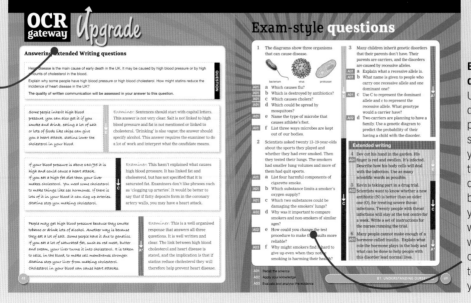

Exam-style questions:

Using these questions you can practice your exam skills, and make sure you're ready for the real thing. Each question has a grade band next to it, so you can understand what level you are working at and focus on where you need to improve to get your target grade.

Routes and assessment

Matching your course

The modules in this book have been written to match the specification for **OCR GCSE Gateway Science Biology B**.

In the diagram below you can see that the modules can be used to study either for **GCSE Biology** or as part of **GCSE Science** and **GCSE Additional Science** courses.

	GCSE Biology	GCSE Chemistry	GCSE Physics
GCSE Science	B1	C1	P1
	B2	C2	P2
GCSE Additional Science	B3	C3	P3
	B4	C4	P4
	B5	C5	P5
	B6	C6	P6

GCSE Biology assessment

The content in the modules of this book matches the different exam papers you will sit as part of your course. The diagram below shows you which modules are included in each exam paper. It also shows you how much of your final mark you will be working towards in each paper.

Unit	Modules tested	%	Type	Time	Marks available
A731	B1 B2 B3	35%	Written exam	1hr 15	75
A732	B4 B5 B6	40%	Written exam	1hr 30	85
A733	Controlled Assessment	25%		4hrs	48

Understanding exam questions

The list below explains some of the common words you will see used in exam questions.

Calculate

Work out a number. You can use your calculator to help you. You may need to use an equation. The question will say if your working must be shown. (Hint: don't confuse with 'Estimate' or 'Predict')

Compare

Write about the similarities and differences between two things.

Describe

Write a detailed answer that covers what happens, when it happens, and where it happens. Talk about facts and characteristics. (Hint: don't confuse with 'Explain')

Discuss

Write about the issues related to a topic. You may need to talk about the opposing sides of a debate, and you may need to show the difference between ideas, opinions, and facts.

Estimate

Suggest an approximate (rough) value, without performing a full calculation or an accurate measurement. Don't just guess – use your knowledge of science to suggest a realistic value. (Hint: don't confuse with 'Calculate' and 'Predict')

Explain

Write a detailed answer that covers how and why a thing happens. Talk about mechanisms and reasons. (Hint: don't confuse with 'Describe')

Evaluate

You will be given some facts, data or other information. Write about the data or facts and provide your own conclusion or opinion on them.

Justify

Give some evidence or write down an explanation to tell the examiner why you gave an answer.

Outline

Give only the key facts of the topic. You may need to set out the steps of a procedure or process – make sure you write down the steps in the correct order.

Predict

Look at some data and suggest a realistic value or outcome. You may use a calculation to help. Don't guess – look at trends in the data and use your knowledge of science. (Hint: don't confuse with 'Calculate' or 'Estimate')

Show

Write down the details, steps or calculations needed to prove an answer that you have been given.

Suggest

Think about what you've learnt and apply it to a new situation or a context. You may not know the answer. Use what you have learnt to suggest sensible answers to the question.

Write down

Give a short answer, without a supporting argument.

Top tips

Always read exam questions carefully, even if you recognise the word used. Look at the information in the question and the number of answer lines to see how much detail the examiner is looking for.

You can use bullet points or a diagram if it helps your answer.

If a number needs units you should include them, unless the units are already given on the answer line.

As part of the assessment for your GCSE Biology course you will undertake a Controlled Assessment task. This section of the book includes information designed to help you understand what Controlled Assessment is, how to prepare for it, and how it will contribute towards your final mark.

Understanding Controlled Assessment

What is Controlled Assessment?

Controlled Assessment has taken the place of coursework for the new 2011 GCSE Sciences specifications. The main difference between coursework and Controlled Assessment is that you will be supervised by your teacher when you carry out some parts of your Controlled Assessment task.

What will I have to do during my Controlled Assessment?

The Controlled Assessment task is designed to see how well you can:

- develop hypotheses
- plan practical ways to test hypotheses
- assess and manage risks during practical work
- collect, process, analyse, and interpret your own data using appropriate technology
- research, process, analyse, and interpret data collected by other people using appropriate technology
- draw conclusions based on evidence
- review your method to see how well it worked
- review the quality of the data.

How do I prepare for my Controlled Assessment?

Throughout your course you will learn how to carry out investigations in a scientific way, and how to analyse and compare data properly. These skills will be covered in all the activities you work on during the course.

In addition, the scientific knowledge and understanding that you develop throughout the course will help you as you analyse information and draw your own conclusions.

How will my Controlled Assessment be structured?

Your Controlled Assessment is a task divided into three parts. You will be introduced to each part of the task by your teacher before you start.

What are the three parts of the Controlled Assessment?

Your Controlled Assessment task will be made up of three parts. These three parts make up an investigation, with each part looking at a different part of the scientific process.

	What skills will be covered in each part?
Part 1	Research and collecting secondary data
Part 2	Planning and collecting primary data
Part 3	Analysis and evaluation

Do I get marks for the way I write?

Yes. In two of the three parts of the Controlled Assessment you will see a pencil symbol (✎). This symbol is also found on your exam papers in questions where marks are given for the way you write.

These marks are awarded for quality of written communication. When your work is marked you will be assessed on:

- how easy your work is to read
- how accurate your spelling, punctuation, and grammar are
- how clear your meaning is
- whether you have presented information in a way that suits the task
- whether you have used a suitable structure and style of writing.

Part 1 – Research and collecting secondary data

At the beginning of your task your teacher will introduce Part 1. They will tell you:

- how much time you have – for Part 1 this should be about 2 hours, either in class or during your homework time
- what the task is about
- about the material you will use in Part 1 of the task
- the conditions you will work under
- your deadline.

The first part of your Controlled Assessment is all about research. You should use the stimulus material for Part 1 to learn about the topic of the task and then start your own research. Whatever you find during your research can be used during later parts of the Controlled Assessment.

Sources, references, and plagiarism

For your research you can use a variety of sources including fieldwork, the Internet, resources from the library, audio, video, and others. Your teacher will be able to give you advice on whether a particular type of source is suitable or not.

For every piece of material you find during your research you must make sure you keep a record of where you found it, and who produced it originally. This is called referencing, and without it you might be accused of trying to pass other people's work off as your own. This is known as plagiarism.

Writing up your research

At the end of Part 1 of the Controlled Assessment you will need to write up your own individual explanation of the method you have used. This should include information on how you carried out your own research and collected your research data.

This write up will be collected in by your teacher and kept. You will get it back when it is time for you to take Part 3.

Part 2 – Planning and collecting primary data

Following Part 1 of your Controlled Assessment task your teacher will introduce Part 2. They will tell you:

- how much time you have – for Part 2 this should be about 2 hours for planning and 1 hour for an experiment
- what the task is about
- about the material you will use in Part 2 of the task
- the conditions you will work under
- your deadline.

Part 2 of the Controlled Assessment is all about planning and carrying out an experiment. You will need to develop your own hypothesis and plan and carry out your experiment in order to test it.

Risk assessment

Part of your planning will need to include a risk assessment for your experiment. To get the maximum number of marks, you will need to make sure you have:

- evaluated all significant risks
- made reasoned judgements to come up with appropriate responses to reduce the risks you identified

- manage all of the risks during the task, making sure that you don't have any accidents and that there is no need for your teacher to come and help you.

Working in groups and writing up alone

You will be allowed to work in groups of no more than three people to develop your plan and carry out the experiment. Even though this work will be done in groups, you need to make sure you have your own individual records of your hypothesis, plan, and results.

This write up will be collected in by your teacher and kept. You will get it back when it is time for you to take Part 3.

Part 3 – Analysis and evaluation

Following Part 2 of your Controlled Assessment task your teacher will introduce Part 3. They will tell you:

- how much time you have – for Part 3 this should be about 2 hours
- what the task is about
- about the answer booklet you will use in Part 3
- the conditions you will work under.

Part 3 of the Controlled Assessment is all about analysing and evaluating the work you carried out in Parts 1 and 2. Your teacher will give you access to the work you produced and handed in for Parts 1 and 2.

For Part 3 you will work under controlled conditions, in a similar way to an exam. It is important that for this part of the task you work alone, without any help from anyone else and without using anyone else's work from Parts 1 and 2.

The Part 3 answer booklet

For Part 3 you will do your work in an answer booklet provided for you. The questions provided for you to respond to in the answer booklet are designed to guide you through this final part of the Controlled Assessment. Using the questions you will need to:

- evaluate your data
- evaluate the methods you used to collect your data
- take any opportunities you have for using mathematical skills and producing useful graphs
- draw a conclusion
- justify your conclusion.

B1 Understanding organisms

Why study this module?

To stay healthy, you need to understand how your body works so you can adopt the behaviours that keep you healthy.

In this module you will learn what you need to eat and how to exercise to stay healthy. You will find out about infectious diseases and how your immune system and medicines can deal with them. You will also find out how your hormones play a key role in your growth and development and in helping your body to function properly.

You will learn how drugs affect your health, and what makes us different from each other.

You should remember

1 You are made of cells which are organised into tissues, organs, and systems.

2 You started as one cell which formed stem cells that specialised to do different jobs.

3 Your skin and stomach acids try to stop microorganisms from entering the body.

4 Your immune system tries to deal with any microorganisms that do enter the body.

5 Healthy eating, exercise, and using medicines wisely can keep you healthy.

6 Some drugs are harmful.

7 Genes in your body control your characteristics; they are inherited from your parents.

8 Genes are found on your chromosomes.

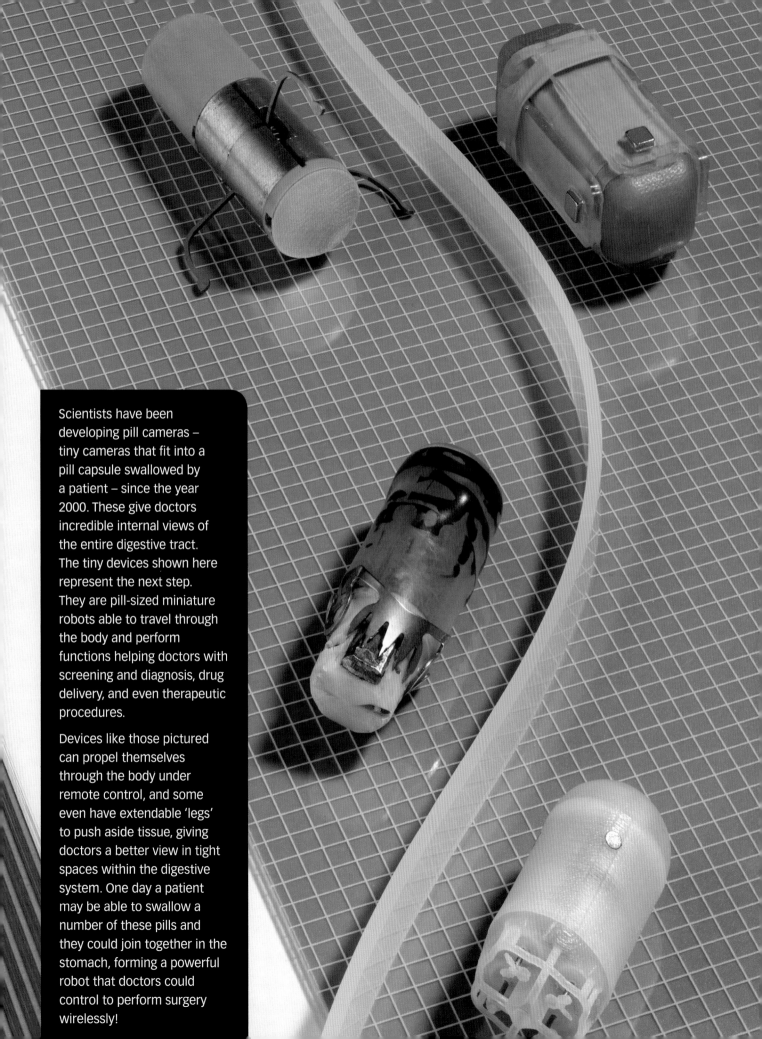

Scientists have been developing pill cameras – tiny cameras that fit into a pill capsule swallowed by a patient – since the year 2000. These give doctors incredible internal views of the entire digestive tract. The tiny devices shown here represent the next step. They are pill-sized miniature robots able to travel through the body and perform functions helping doctors with screening and diagnosis, drug delivery, and even therapeutic procedures.

Devices like those pictured can propel themselves through the body under remote control, and some even have extendable 'legs' to push aside tissue, giving doctors a better view in tight spaces within the digestive system. One day a patient may be able to swallow a number of these pills and they could join together in the stomach, forming a powerful robot that doctors could control to perform surgery wirelessly!

Learning objectives

After studying this topic, you should be able to:

- ✔ evaluate the different ways of measuring fitness
- ✔ recall that blood in arteries is under pressure, and this can be measured
- ✔ describe the factors that affect blood pressure
- ✔ explain the difference between fitness and health
- ✔ explain the link between saturated fat, cholesterol, and heart disease

Key words

disease, systolic pressure, diastolic pressure, thrombosis

▲ Strength training

Exam tip | OCR

- ✔ Do not refer to cholesterol 'clogging up' arteries. Use the technical terms 'thrombosis' or 'blood clot'.

What is physical fitness?

Physical fitness is the ability to do physical activity.

You can measure your fitness level by measuring your
- stamina (endurance)
- strength
- flexibility
- agility (being able to move quickly and nimbly)
- speed
- cardiovascular efficiency (an efficient heart and normal blood pressure).

Each of the above methods measures a different aspect of physical fitness. Some methods are more difficult to carry out as they may need specialist equipment.

What is good health?

Good health is being free from **disease**. A disease is a condition caused by any part of the body not functioning properly.

> A What is physical fitness?
> B What is good health?

Heart disease

The main cause of early death in the UK is heart disease. Your risk of developing heart disease is increased if you
- have high blood pressure
- eat too much saturated fat
- smoke
- eat too much salt.

Blood pressure

Each time the heart beats its muscles contract. This pushes the blood out into the arteries at the correct pressure so it reaches all parts of the body. The two blood pressure measurements, such as 120/80 mmHg, show
- the pressure in the arteries when the heart contracts (**systolic pressure**) (120 mmHg)
- the pressure in the arteries when the heart relaxes (**diastolic pressure**) (80 mmHg).

During exercise your systolic pressure goes up, but then comes down again when you have finished exercising.

High blood pressure

The following can increase your resting blood pressure:

- smoking
- eating too much salt
- being overweight
- stress
- drinking too much alcohol regularly
- eating a lot of saturated fat.

Regular exercise and eating a balanced diet help lower your resting blood pressure to normal.

If you have high blood pressure (above 140/90 mmHg) while you are resting you may be more likely to
- have a heart attack
- have a stroke (or a blood vessel in the brain may burst)
- suffer from kidney damage.

How smoking increases your blood pressure

Tobacco smoke contains carbon monoxide and nicotine.

- Carbon monoxide combines with haemoglobin in red blood cells and prevents them carrying as much oxygen. The heart beats faster to compensate. This puts a strain on the heart.

- Nicotine increases the heart rate.

How saturated fat may increase blood pressure

- Your liver makes cholesterol from saturated fat (fat in milk, butter, eggs, cheese, red meat, cream).
- Cholesterol is carried in the blood and may be deposited in artery walls, forming fatty plaques.
- These deposits narrow the arteries and restrict blood flow. The blood pressure increases to force blood through the narrower gap.

Low blood pressure

Having resting blood pressure that is too low can also be harmful. It can lead to
- dizziness
- fainting
- poor circulation and organ failure.

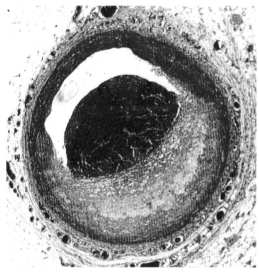

▲ An artery with a fatty deposit

Thrombosis

Cholesterol deposits may also lead to a blood clot (**thrombosis**). A thrombosis in the artery supplying the heart muscle causes a heart attack. A thrombosis in an artery supplying the brain causes a stroke.

Questions

1 How is blood transported to all parts of the body at the correct pressure?

2 A person has a blood pressure measurement of 130/85 mmHg. Explain what these figures mean.

3 Name two things that can lower resting blood pressure.

4 Explain how smoking causes high blood pressure.

5 Explain how eating too much saturated fat may increase your blood pressure.

6 State three health risks from having high blood pressure.

E

↓ C

A*

2: Health and diet

Learning objectives

After studying this topic, you should be able to:

✔ know what makes a balanced diet
✔ understand the difference between physical and chemical digestion

Key words

balanced diet, deficiency disease

▲ The largest part of a healthy, balanced diet should be carbohydrates

A What is meant by a balanced diet?

B Which parts of a balanced diet give you energy?

EARs are averages

Remember that the EAR is an average amount. However, there is really no such thing as 'the average person'. A person's protein needs may vary depending on age, pregnancy or lactation (making milk).

What is a balanced or healthy diet?

Like other animals, you have to get food by eating it. Food is a source of energy. A **balanced diet** contains the right amount of the different foods and the right amount of energy to keep you healthy. It also includes nutrients that do not provide energy: minerals and vitamins, water, and fibre.

Food/nutrient	Why you need to eat it	Where it is stored
carbohydrates (made up of simple sugars like glucose)	to give you energy	stored in the liver as glycogen or converted to fats
fat (made up of fatty acids and glycerol)	to give you energy	stored under the skin and around organs as adipose tissue
proteins (made up of amino acids)	for growth and repair (you can use proteins for energy if you don't have enough fat or carbohydrate)	not stored
iron (a mineral)	to make haemoglobin	
vitamin C (a vitamin)	to prevent scurvy	
fibre	to prevent constipation	
water (your cells contain about 70% water)	to prevent dehydration, replacing water lost in tears, urine, and faeces	

Some people do not eat a balanced diet. They eat the wrong amount or types of food. A balanced diet is not the same for everyone. It may vary according to:

- age
- religion
- medical reasons (eg diabetes, allergies)
- gender
- how active you are
- personal choice (eg vegetarian/vegan)

We need to eat enough protein
Calculating the EAR for protein

We can calculate how much protein we should eat each day. This is called the estimated average requirement or EAR.

$$\text{EAR in g} = 0.6 \times \text{body mass in kg}$$

So a 60 kg boy should eat $0.6 \times 60 = 36$ g of protein daily.

What happens if you do not eat enough protein?

You are growing fast, and need protein to grow and develop properly.

In some developing countries people do not get enough protein. This is linked to

- overpopulation – too many people to feed properly
- limited investment in agricultural techniques which results in less food being produced

> - eating more second class proteins (from plants) than first class proteins (from animals). Most animal proteins provide all the essential amino acids, whereas individual plant proteins do not. Given the wide variety of plant/vegetable protein sources that can be consumed within the diet, the best advice for those not eating meat is to consume a wide variety of different plant proteins.

If children do not eat enough protein they suffer from a **deficiency disease** called kwashiorkor. They have very thin arms and legs and a swollen belly. Their bodies can't fight infection very well.

In developed countries some people choose to eat too little. They may have low self-esteem and a poor self-image and think that by being thin they will look better. Being so undernourished can lead to heart problems and poor health.

BMI

You can work out whether you have a healthy weight by calculating your body mass index or BMI. This is your mass in kg divided by your height in metres, squared.

So if a girl weighs 50 kg and is 1.60 m tall, her BMI is $\frac{50}{(1.6)^2}$ which is $\frac{50}{2.56} = 19.53$.

A person with a BMI of

- under 18 is underweight
- 18 to 25 is normal
- 25 to 30 is overweight
- over 30 is obese.

You have seen that there are health risks linked to being underweight. If you are very overweight there are also health risks. Obese people are more likely to suffer from

- heart disease
- diabetes
- arthritis
- breast cancer.

C Calculate the EAR for protein for an 80 kg man.

Exam tip **OCR**

✓ For a calculation, always show your working. Then if you use the right method but make a mistake, you will get some marks for method. If you just write down a wrong answer, you will get no marks.

Questions

1 Why do you need to eat (a) vitamins (b) minerals (c) fibre (d) water? ↓E

2 What are the health risks of being obese?

3 (a) Calculate the BMI for a woman who is 1.5 m tall and weighs 70 kg. (b) Is she normal, underweight, overweight, or obese? ↓C

4 Why do you think vegetarians eat milk and eggs, although these come from animals? Share your ideas with the rest of your class. ↓A*

Learning objectives

After studying this topic, you should be able to:

- ✔ describe the difference between infectious and non-infectious diseases
- ✔ recall the meaning of the terms parasite and host
- ✔ state that some diseases can be caused by dietary deficiency, body disorders, and genetic inheritance
- ✔ describe lifestyle changes that may reduce the risk of some cancers

Key words

microorganism, parasite, host, cancer, tumour, **benign**, **malignant**

Did you know...?

Although cancer has usually been described as non-infectious, some types of cancer are infectious. They are caused by viruses. Cervical cancer is caused by a virus and so there is now a vaccine against it.

Infectious and non-infectious diseases

Infectious diseases are caused by **microorganisms** that invade your body. These harmful microorganisms are **parasites**. They live in your body and gain nutrients and shelter. Your body is the **host** for the parasites. You can read more about infectious diseases in the next spread.

Non-infectious diseases are not caused by a parasite invading your body. There are many causes of non-infectious diseases. The table shows some of them.

Cause of non-infectious disease	Example(s)
vitamin deficiency	scurvy (lack of vitamin C)
mineral deficiency	anaemia (lack of iron)
body disorder	diabetes; cancer
genetic inheritance	red-green colour blindness

A What causes infectious diseases?

B Describe the difference between infectious and non-infectious diseases.

C State four different types of non-infectious disease.

D How would you avoid developing anaemia?

Cancer

There are many types of **cancer**. In cancer, cells in the body keep on dividing and form an abnormal mass called a **tumour**.

Many cancers are not infectious. You can reduce your risk of certain cancers by adopting a healthy diet and lifestyle. For example, if you do not smoke you are unlikely to get lung cancer. To reduce their risk of cancer people should also

- sunbathe a little, but make sure they do not burn
- avoid eating too much fat and make sure they do not become overweight
- avoid eating too much red meat and processed food
- eat plenty of fresh fruit and vegetables
- take regular exercise
- avoid drinking too much alcohol.

(a)

(b)

(a) A less healthy diet contains no fruit and vegetables and a lot of saturated fat.
(b) A healthy diet contains colourful fruit and vegetables and is low in fat.
Eating a healthy diet has many health benefits, including reducing your risk of cancer.

Tumours

People with cancer have a tumour. If the tumour does not spread to other parts of the body, it is **benign**. If it spreads to other parts of the body, it is **malignant**. A benign tumour is easier to deal with as it can be cut out. Once a tumour spreads there may be many cancer cells all over the body.

Questions

1   List five ways to reduce your risk of getting cancer.

2   Explain the terms parasite and host.

3   Describe the difference between a benign tumour and a malignant tumour.

4   The graph shows survival rates for men diagnosed with a type of cancer.

   (a) What percentage survived for 10 years after diagnosis in 1986–1990?

   (b) What percentage survived for 10 years after diagnosis in 1991–1995?

   (c) What do you think are the reasons for the change in survival rate?

C

A*

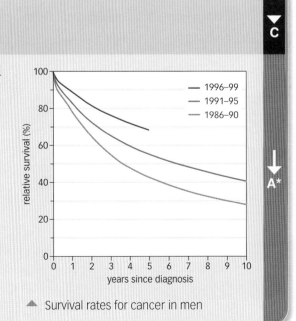

Survival rates for cancer in men

4: Infectious diseases and antibiotics

Learning objectives

After studying this topic, you should be able to:

- ✔ state that infectious diseases are caused by pathogens
- ✔ recall types of pathogen, such as fungi, bacteria, viruses, and protozoa
- ✔ describe how vectors spread disease
- ✔ interpret data on the incidence of disease around the world
- ✔ recall that antibiotics can be used to treat bacterial and fungal infections
- ✔ recall the difference between antibiotics and antivirals

A What causes infectious diseases?

B How do pathogens make us ill when they invade our bodies?

C State four types of pathogen and name one disease caused by each type.

D The malaria parasite is spread by mosquitoes. Name another host for the malaria parasite.

▲ This mosquito may spread the malaria parasite from one person to another

What causes infectious diseases?

Infectious diseases are caused by parasitic microorganisms that infect you. These disease-causing microorganisms are called **pathogens**. The invading pathogens may damage cells and also release chemicals called **toxins**.

Type of pathogen	Example of disease caused
fungi	athlete's foot
bacteria	cholera
viruses	flu
protozoa	malaria

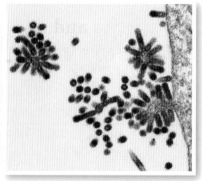

▲ Bacteria. Some are rod shaped (shown here as grey), some are round (red), and some are spiral (cream).

▲ Electron micrograph of influenza viruses (× 75 000). These have reproduced inside a cell of the respiratory tract and are breaking out, causing damage.

Infectious diseases spread

Some infectious diseases are spread from person to person by **vectors**. Malaria is caused by a parasite which is spread by female mosquitoes. When the mosquitoes bite people they can spread the disease. Humans and mosquitoes are both hosts for the malaria parasite.

Knowing how the vector spreads the disease allows people to take steps to control malaria. These include

- sleeping under a mosquito net and using insect repellent
- draining areas of stagnant water
- using insecticides to kill the mosquitoes.

Incidence of disease

The **incidence** of a disease is the rate at which new cases occur in a population each year. This is usually given as the number of cases per 10 000 people or per 100 000 people. The incidence of a particular disease can depend on climate and socio-economic factors.

Climate

In places where it is always warm or hot, vectors may multiply rapidly. The malaria mosquito needs water to breed, so a high rainfall also increases the incidence of malaria.

Socio-economic factors

The incidence of infectious diseases such as cholera is high in countries where there is no clean drinking water or proper sewage system.

Antibiotics and antivirals

Antibiotics are chemicals produced by some fungi and bacteria, to kill or prevent the growth of other bacteria or fungi. We use them to kill pathogens in our bodies when we have an infectious disease. Penicillin was the first antibiotic to be discovered, in the 1920s. It has been used to treat people since the 1940s. Since then many others have been found.

Antibiotics do not deal with viruses because viruses do not grow and have no metabolic reactions to be prevented. Antiviral drugs, such as Tamiflu and Relenza, inhibit the replication of viruses inside the host.

Antibiotics have to be used carefully. Doctors should not prescribe them if they are not needed. Patients taking antibiotics must finish the whole course, even when they begin to feel better. If antibiotics are not used carefully, then resistant strains of bacteria develop. MRSA is an example. This strain of a bacterium lives on our skin and does no harm. However, if it gets into a wound after surgery it causes a bad infection. It is resistant to many antibiotics and so it is difficult to deal with. However, it can be killed by antiseptics, so cleaning your skin thoroughly before an operation prevents the infection. Iodine solution is a good antiseptic for this.

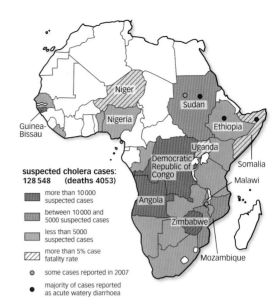

suspected cholera cases:
128 548 (deaths 4053)

▨ more than 10 000 suspected cases

▨ between 10 000 and 5000 suspected cases

▨ less than 5000 suspected cases

▨ more than 5% case fatality rate

◉ some cases reported in 2007

● majority of cases reported as acute watery diarrhoea

▲ Climate and socio-economic factors influence the incidence of disease around the world. This map shows the incidence of cholera in February 2009 in African countries.

Key words

pathogen, toxin, vector, incidence, antibiotic

Questions

1 The incidence of malaria is high in areas where it is warm and wet, like the tropics and subtropics. In these areas many people are poor and cannot afford a mosquito net to sleep under. Name two climate factors and one socio-economic factor that link to malaria. **↓ E**

2 What are antibiotics?

3 Which was the first antibiotic to be discovered? **↓ C**

4 What is MRSA?

5 Explain why antibiotics have to be used carefully. **↓ A★**

> **A** How does your body stop pathogens entering it?

Did you know...?

Your immune system also protects you from cancer. It can recognise and kill cancerous cells.

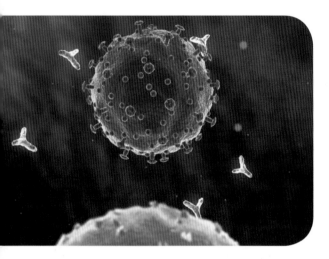

▲ Antibodies surrounding a virus particle

> **B** Describe two ways that white blood cells can kill pathogens.
>
> **C** Explain how you become immune to a disease such as measles.

How your immune system deals with pathogens

Your body has barriers to stop pathogens entering it. These include

- skin
- stomach (hydrochloric) acid
- mucus in the airways
- tears
- blood clotting when you cut yourself.

However, sometimes some pathogens manage to enter your body. When they do your **immune system** may deal with them. This involves your white blood cells.

White blood cells

You have different sorts of white blood cells.

- Phagocytes engulf (ingest) the pathogens.
- Lymphocytes produce antibodies or antitoxins.

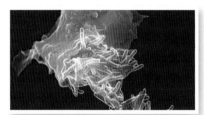

▲ Electron micrograph of a phagocyte ingesting TB bacteria (×3500)

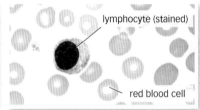

lymphocyte (stained)

red blood cell

▲ Light micrograph of lymphocytes, the white blood cells that make antibodies (×400)

Antibodies

Antibodies are proteins. Each type of antibody can destroy a particular type of bacterium or virus. This is because:

- Each type of pathogen has particular **antigens** (proteins) with a specific shape on its surface.
- Each type of antibody (also a protein) has a particular shape and can lock onto a particular antigen.
- Your immune system makes the right sort of antibodies to lock onto the antigens of the particular pathogen that is in your body.
- Once the pathogen is coated with antibodies, white blood cells can ingest and kill the pathogens.

Once you have recovered from this infection, you have **immunity** to it. If that same pathogen enters your body again your body makes antibodies so quickly that the pathogen is destroyed before it makes you feel ill.

Immunisation

Many people used to die from infections like TB and smallpox. In the developing world today many die from measles, malaria, and cholera. So it is better not to get these infections in the first place. **Immunisation** can make people immune to a disease, without them having the disease.

Here's how it works:

- A small amount of dead or inactivated pathogen is introduced into your body – usually by injection. This is called being vaccinated. The dead or inactivated pathogens still have the antigens on their surface.
- Some of your white blood cells recognise these antigens on the pathogens and respond to them by making antibodies. Other white cells become memory cells.

> - If, later on, the live pathogens get into your body, these memory cells quickly make the right sort of antibodies.
> - These antibodies destroy the pathogens before they make you ill.

How do mutations affect vaccines?

Some viruses, like the flu virus, mutate often. This causes them to have a slightly different structure. Your immune system does not recognise these viruses so they can make you ill again with flu, even though you may have had flu before. So every year new vaccines are made for the new strains of flu that are likely to infect people that year.

In 2009 many people in many countries were immunised against bird flu to prevent the disease sweeping across countries and causing an **epidemic**, or across continents and causing a **pandemic**.

Active and passive immunity

You have seen that you can become immune after having an infection, or after having a vaccination. Both of these are types of **active immunity** because you make your own antibodies.

Sometimes you may be injected with antibodies made in a lab. This is called **passive immunity** because you do not make your own antibodies. This is useful if you have been infected by a pathogen that can quickly kill you. However, passive immunity does not last long, unlike active immunity.

D Draw a flow diagram to show how immunisation works.

Exam tip

- ✓ Do not confuse the terms 'immunity' and 'resistance'. People become immune to infectious diseases because they have an immune system. Bacteria (not people) may be resistant to antibiotics.

Questions

1 What is (a) an antigen (b) an antibody?

2 Describe one way that pathogens can enter your body.

3 Describe two ways that you can become immune to mumps.

4 Explain how mutations in viruses can lead to epidemics.

5 Healthcare workers are vaccinated each year against flu. Why do you think this is?

Learning objectives

After studying this topic, you should be able to:

✔ name the main parts of the eye
✔ describe how light passes through the eye
✔ describe monocular and binocular vision
✔ state the main problems of vision and their causes

Did you know...?

Eyes detect light, but you really see with your brain. This is because it is the brain that interprets the image formed at the back of your eye.

A State the functions of the following structures in the eye: iris, pupil, cornea, lens, retina, optic nerve.

B What is the blind spot?

▲ The zebra's eyes are on each side of its head. It can see all around and notice any predators.

The structure and function of the eye

Eyes contain receptors that detect light. These receptors change light into electrical impulses. The impulses pass along the optic nerve to the brain.

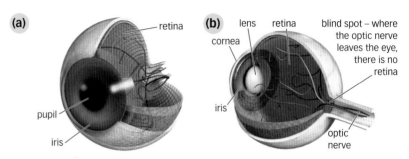

▲ Structure of the eye (a) seen from the front and (b) seen from the back

Binocular and monocular vision

Animals with both eyes facing forward have **binocular vision**. Both eyes **focus** on the same thing. This enables them to focus well and to judge distance accurately. This is very useful for predators when they are hunting. However, they do not have a very wide field of view.

Animals with an eye on each side of the head have **monocular vision**. This gives them a much wider field of view. They can see all around and notice any predators in time to run away. However, they cannot judge distances very well. Humans have binocular vision. This gives them a narrower field of view but better judgement of distance.

To judge distance, your brain compares the images from each eye. If the images are very similar, the brain knows you are looking at something at a distance.

Humans can focus on near or distant objects, but not both at the same time. The lens has to change shape. This is called **accommodation**.

Structure	Accommodation for close viewing	Accommodation for distant viewing
ring of ciliary muscle	contracted	relaxed
suspensory ligaments	slacken	become taut
lens	rounded, fat shape	flatter, thinner shape

Problems with vision

Red-green colour blindness is an inherited condition. People cannot clearly tell the difference between red and green colours. They do not have certain specialised cells in the retina.

People with **long sight** can see things in the distance clearly, but cannot see close work clearly.

People with **short sight** can see close things clearly, but cannot focus on things in the distance.

C Describe the difference between monocular and binocular vision.

D What are the advantages of binocular vision?

E What are the disadvantages of binocular vision?

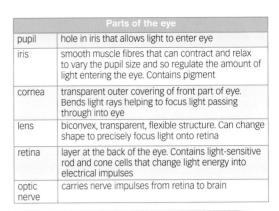

Parts of the eye	
pupil	hole in iris that allows light to enter eye
iris	smooth muscle fibres that can contract and relax to vary the pupil size and so regulate the amount of light entering the eye. Contains pigment
cornea	transparent outer covering of front part of eye. Bends light rays helping to focus light passing through into eye
lens	biconvex, transparent, flexible structure. Can change shape to precisely focus light onto retina
retina	layer at the back of the eye. Contains light-sensitive rod and cone cells that change light energy into electrical impulses
optic nerve	carries nerve impulses from retina to brain

Key words

binocular vision, focus, monocular vision, accommodation, red-green colour blindness, long sight, short sight

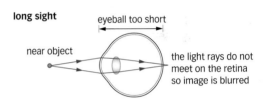

▲ In short sight, the eyeball is too long

▲ In long sight, the eyeball is too short

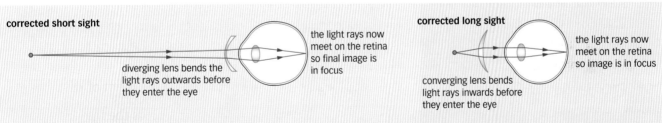

▲ Spectacles or contact lenses with concave lenses are used to correct short sight. These bend the light rays outwards, helping the eye to focus light on the retina. For long sight, convex lenses bend light rays inwards. This makes the light focus on the retina. Surgery on the cornea can also be used to correct sight.

Questions

1 What causes red-green colour blindness?

2 Explain the difference between long sight and short sight.

3 What is accommodation in vision?

4 Describe how long sight and short sight can be corrected.

Exam tip   OCR

✔ Remember that concave lenses bend light rays outwards and are used to correct short sight. Convex lenses bend light rays inwards and are used to correct long sight.

Learning objectives

After studying this topic, you should be able to:

✔ understand the role and organisation of the nervous system

✔ know that receptors detect stimuli

✔ know that nerve impulses pass from receptors along neurones

✔ recall how reflex actions come about

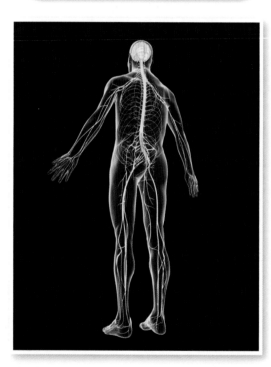

▲ The nervous system is made up of the central nervous system (CNS) and the peripheral nervous system

Key words

stimulus, central nervous system, peripheral nervous system, receptor, neurone, reflex action, response, synapse, effector

Why you need to respond to change

You, and all living things, need to be able to respond to changes in the environment. These changes are called **stimuli**. If you could not detect and respond to stimuli you would not be able to find food or avoid danger. You also need to learn from your experiences and coordinate your behaviour.

The structure of the nervous system

There are two main parts to the nervous system:

* The **central nervous system** (CNS) – the brain and spinal cord.
* The **peripheral nervous system** – nerves taking information from sense organs into the CNS, and from the CNS to effectors (muscle or glands).

A What is a stimulus?

B Name the two parts of the nervous system.

C Why do we need to be able to detect stimuli?

Sense organs or receptors

Receptor cells are special cells adapted to detect stimuli. Like most animal cells they have a nucleus, a cell membrane, and cytoplasm.

Sense organ	Information
skin	pressure, temperature, touch, pain
tongue	chemicals in food (taste)
nose	chemicals in air (smell)
eyes	light (sight)
ears	sound (hearing), balance

▲ The body's sense organs (receptors) and the information they gather

Information from the receptors passes as electrical impulses. It travels along nerve cells called **neurones** to the brain. The brain then coordinates the response. Some responses are voluntary – they are consciously controlled by the brain. For example you may hear part of a song on the radio and decide whether to turn up the volume or switch off the radio.

Reflex actions

Sometimes you need to respond to a potentially dangerous situation very quickly. For example, if you touch a hot object, you need to quickly withdraw your hand before it burns. There is no time to think, so the brain does not need to be involved. Instead the response is coordinated by the other part of the CNS, the spinal cord. These responses are called (spinal) **reflex actions**. They are fast, automatic, and protective.

Each reflex action follows the pathway: stimulus → receptor → sensory neurone → relay neurone → motor neurone → effector → response.

This pathway is described as a reflex arc.

▲ The knee-jerk reflex test. When the leg is tapped just below the knee, the leg straightens. This reflex is used when we walk.

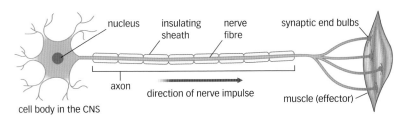

3 A nerve impulse travels along the sensory neurone.

4 The nerve impulse enters the spinal cord.

5 The nerve impulse passes across a synapse into a relay neurone.

2 Pain receptors in the skin are stimulated.

spinal cord

1 The stimulus – a pin prick.

7 When the nerve impulse reaches the finger muscle (the effector), the muscle contracts, pulling the finger away from the pin – the **response**.

6 The nerve impulse passes into a motor neurone and travels along this out of the spinal cord.

▲ A reflex arc. The impulse goes from receptor to CNS and then to effector to bring about the response. The relay neurone (purple) inside the spinal cord coordinates the response by connecting the sensory neurone to an appropriate motor neurone (green). The information travels from one neurone to another across a small gap called a **synapse**.

Neurones are adapted to their functions because they are long, have an insulating sheath to prevent impulses leaking away, and have branched endings so they can communicate with many other neurones.

Gaps between neurones are called synapses. Electrical nerve impulses cannot cross these gaps. Instead, chemicals called neurotransmitters are released by one neurone, diffuse across the synapse and fit onto receptors on the next neurone, causing an impulse to be fired in it.

Effectors

Effectors are glands or muscles that carry out a response.
- A muscle responds by contracting.
- A gland responds by secreting chemical substances.

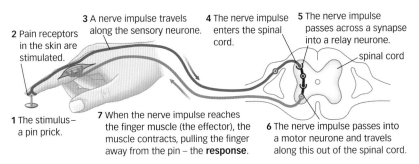

nucleus   insulating sheath   nerve fibre   synaptic end bulbs

axon   direction of nerve impulse   muscle (effector)

cell body in the CNS

▲ Structure of a motor neurone. The nerve impulse is carried along the nerve fibre (axon).

Examples of reflex actions include
- the knee-jerk reflex
- withdrawing a hand from a hot plate.

Questions

1   What are voluntary responses?

2   Why are reflexes automatic and rapid?

3   How does the pupil-shrinking reflex protect the eye?

↓ E

4   Which part of the neurone does the impulse travel along?

↓ C

5   What is the function of the insulating sheath?

↓ A*

Key words

drug, tolerant, addiction, withdrawal symptoms, rehabilitation, **blind trial**, **double blind trial**

🔺 Opium is obtained from the seed heads of opium poppies. It contains morphine and codeine, and can also be refined to make the illegal drug heroin.

Did you know...?

Some animals self-medicate. They eat certain leaves that they do not normally eat, to treat parasitic infections.

A Name three drugs that can be obtained from opium poppies.

B Why do you need to have a prescription from a doctor for certain drugs, like strong painkillers?

Drugs may be beneficial or harmful

A **drug** is a chemical that alters the way your body or brain works. Drugs may alter your behaviour as well as altering your metabolism.

Beneficial drugs are medicines like painkillers (morphine and codeine) and antibiotics. Some have to be prescribed by a doctor. This is because they may

- have side effects
- interfere with another medicine the patient is taking
- be harmful for a particular patient if they have another condition
- be harmful if taken too often.

Some drugs are legal and used for recreation. These include caffeine, nicotine in tobacco, and alcohol.

Some athletes use performance-enhancing drugs like anabolic steroids. These can have harmful side effects. It may be unethical to use them as it gives some athletes an unfair advantage. The athletes may also suffer side effects from taking the anabolic steroids.

Properties of drugs

Because all drugs alter your body chemistry, your body may become **tolerant** to them. You may need to increase the dose to get the same effects.

Eventually your body may not work properly without the drug. Then you are **addicted**. Heroin and cocaine are very addictive.

If you stop taking the drug you get **withdrawal symptoms**. Depending on the drug, these can include

- nausea
- diarrhoea
- hallucinations
- vomiting
- shaking
- craving for the drug.

Rehabilitation

People who want to 'kick the habit' need **rehabilitation**. They may go into a hospital or special clinic. After not taking the drug for a few days, withdrawal symptoms fade and they feel better. They can cope without the drug. The staff in the clinic help and support the patients. They also prevent them taking the drug.

Types of drug

Type of drug	Example(s)	Effects on your body and mind
depressants	alcohol, solvents, temazepam	slow down brain and nerve activity
painkillers	aspirin, paracetamol	block nerve impulses
stimulants	nicotine, caffeine, ecstasy	increase brain activity and counteract depression
performance enhancers	anabolic steroids	increase muscle development
hallucinogens	cannabis and LSD	distort what you see/hear

How depressants and stimulants work

Depressants reduce activity at synapses. They bind to the receptor molecules on the next neurone and block transmission of the nerve impulse. Stimulants, on the other hand, cause more transmitter substance to cross the synapse.

Illegal drugs

Some people use illegal drugs for recreation. Some drugs are more harmful than others. The Government has introduced the Misuse of Drugs Act. This divides drugs into categories.

Class		Examples	Penalties for possession	Penalties for dealing
A	most dangerous, heaviest penalties	cocaine, heroin, LSD, mescalin, ecstasy, some types of cannabis	up to 7 years in prison, unlimited fine	up to life imprisonment, unlimited fine
B		amphetamines, cannabis, strong codeine, ritalin, barbiturates, mephedrone	up to 5 years in prison, unlimited fine	up to 14 years in prison, unlimited fine
C	least dangerous, lightest penalties	temazepam, anabolic steroids	up to 2 years in prison, unlimited fine	up to 14 years in prison, unlimited fine

Testing new drugs

New drugs are tested before use to see if they are toxic or have side-effects, and to find the lowest effective dose. Tests use computer models then mammalian tissue and laboratory animals. They are then trialled on human volunteers. Some people object to drugs being tested on animals for ethical reasons and because the effects on humans may be different. Testing is regulated in the UK.

Randomised controlled clinical trials are used to see whether new drugs are better than existing drugs or a placebo (dummy pill). In a **blind trial** the groups do not know who is receiving the new drug, the existing drug or placebo. In a **double blind trial** neither the volunteers nor the doctors collecting data know. This is to avoid experimental bias. Sometimes people feel better because they think they are receiving a drug (placebo effect).

Questions

1. Name two types of class B drugs.
2. Why do you think new drugs have to be tested before they are licensed for use as medicines?
3. What are hallucinogens?
4. Explain how (a) depressants (b) stimulants work.

Learning objectives

After studying this topic, you should be able to:

✔ describe the effects of carbon monoxide, nicotine, and tar on the body

✔ describe the effects of alcohol on the body

▲ Tobacco plant, *Nicotiana tabacum*. This plant is native to tropical America. It produces the toxin nicotine, which protects it from insects and grazing animals. Small doses of nicotine produce a pleasurable stimulus in humans but habitual use leads to addiction. It also leads to health problems caused by other chemicals in tobacco smoke.

▲ Tobacco farming. Tobacco leaves are dried in well ventilated barns for 4–8 weeks.

Key words

carcinogen, emphysema, addictive, depressant

Tobacco

Tobacco comes from leaves of the tobacco plant, *Nicotiana*. It was used by native Americans. They smoked it in pipes in religious ceremonies and when making a bargain. When Europeans arrived in the Americas in the 1500s, they used it as a recreational drug and for trade. Cigarettes were mass produced from the early 1900s and many people began to smoke them.

A burning cigarette produces carbon monoxide, nicotine, tar, and particulates in the smoke. Lung cancer takes about 30 years to develop.

What do the substances in tobacco do to you?

When you smoke a cigarette, the smoke passes into your lungs. Substances in the smoke then get into your blood and are carried around the body.

Substance	Effect on your body
carbon monoxide	Reduces the oxygen-carrying capacity of the blood so less oxygen is taken to cells and tissues. The heart has to work harder and this contributes to heart disease. If a pregnant woman smokes, less oxygen gets to the fetus. This leads to low birth mass in the baby.
tar	Is an irritant. Is **carcinogenic** (causes cancer) of lungs, throat, oesophagus, mouth, and bladder. May also damage the ciliated epithelial cells in the trachea, bronchi, and bronchioles, and prevent the cilia from working properly. Mucus accumulates, which may cause a smoker's cough or bronchitis.
particulates (small particles)	These accumulate in lung tissue. White blood cells ingest them and release an enzyme that harms alveoli and leads to **emphysema**.
nicotine	This is **addictive**. Smokers become tolerant and dependent making it hard to stop. It also increases heart rate and puts strain on the heart.

A How is tobacco made?

B How do humans use tobacco as a recreational drug?

C For each of the following component of tobacco smoke, describe how it affects the body: nicotine; carbon monoxide; tar.

Alcohol

Alcohol has been drunk by humans since prehistoric times.

- It can make people feel relaxed as it depresses the activity of the nervous system.
- If you drink it in small amounts it can be good for your health.
- But it is a **depressant** drug and you develop tolerance to it. Some people become dependent on it (addiction).
- Although in many cultures alcohol is a legal drug, drinking too much can cause a lot of harm.
- Binge drinking is especially harmful.

Alcohol has short-term and long-term effects on your body.

Short-term effects of alcohol	Long-term effects of alcohol
• Blurred vision • Impaired speech • Impaired judgement and mental confusion • Increased reaction times • Impaired balance and muscle control • Loss of self-control • Violent behaviour • Drowsiness • Unconsciousness or coma • Increased blood flow to the skin which means you feel warm but are losing core body heat	• Cirrhosis of the liver. Enzymes in liver cells break down the toxic alcohol. This produces other toxic products that damage the liver. • Brain damage including loss of memory and depression • Can damage the fetus if a pregnant woman drinks • Weight gain • Fatty liver (beer gut) • Increased risk of diabetes and heart disease • Cancer of liver, mouth, throat, and oesophagus

The legal limits for drivers and pilots

Because alcohol impairs judgement and slows reactions, drivers and pilots cannot perform properly if they have drunk too much alcohol. Their risk of having an accident is greatly increased. So there is a legal limit for the amount of alcohol allowed in their blood and breath.

It is difficult to work out exactly how much a person can drink if driving, as everyone is different. As a general rule, drivers should only drink two units of alcohol. However, people tend to think they are more capable than they actually are after drinking. It is probably best not to drink any alcohol if driving.

Campaigns to discourage drinking and driving have led to a reduction in the number of road casualties.

▲ One unit of alcohol is half a pint of normal strength beer; one small glass (125 ml) of wine; one small glass (50 ml) of sherry or port or one measure (25 ml) of spirits (vodka, gin, whisky, brandy).

Did you know...?

Tobacco plants are in the same family as deadly nightshade. The toxin nicotine was named after Jean Nicot who, in 1559, sent tobacco as a medicine to the French court.

Exam tip

✔ You may be asked questions that require you to interpret information about accident statistics and alcohol level. Always read the graphs or tables carefully.

Questions

1. State six short-term effects of drinking alcohol.
2. State four long-term effects of drinking alcohol.
3. Explain why it is advisable for pilots and drivers not to drink alcohol.
4. Why does alcohol cause liver damage?

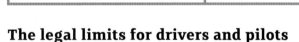

Key words

homeostasis, **vasoconstriction**, **vasodilation**, evaporation

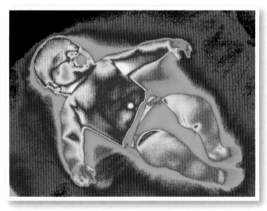

▲ Thermogram of a baby. The hottest areas are white. The temperature scale then goes red, yellow, green, blue, and purple (coldest).

A Which parts of the baby are hottest?

B Which parts of the baby are coolest?

C Are the baby's internal organs kept warm or cool?

Staying in balance

Inside all the cells of your body, lots of chemical reactions are going on, all controlled by enzymes. Your cells will not function well if your body

- is the wrong temperature
- does not have the right amount of water
- has too much carbon dioxide in the blood.

This is because the enzymes cannot work properly.

To maintain a constant internal environment, inputs and outputs need to be balanced in the body. This balancing is called **homeostasis**.

The factors shown above are kept at steady levels by automatic control systems involving nerves and hormones.

Body temperature

The temperature of the human body is normally kept at about 37°C. This is the core temperature. At your fingers and toes it will be cooler.

How and where body temperature is measured

You can measure your body temperature with
- a clinical thermometer, in the ear, mouth, or anus
- temperature-sensitive strips on the forehead
- digital recording probes on a finger, linked to a computer.

Scientists and doctors can also use thermal imaging. The heat given off by parts of the body is converted to a visual image. This shows which areas of the body are warmest and coolest.

Too hot or too cold

It is dangerous if your body temperature goes above 40°C or below 35°C:
- Above 40°C you will get heat stroke and dehydration. If this is not treated you may die.
- Below 35°C your body develops hypothermia and if not treated this also leads to death.

How your body gains and loses heat to keep the core temperature constant

If your internal organs start to get too hot, the blood flowing through them carries the excess heat away. The increased temperature is monitored by the brain. The brain then brings about mechanisms to control temperature.

How the body gains heat in cold conditions	How the body loses heat in hot conditions
• Respiration in cells releases some energy as heat. • Shivering (muscles contracting) generates heat. • Less blood flows near the skin surface (**vasoconstriction**). • Exercise generates heat because muscles need to respire more to contract more. • Wearing more clothes insulates the body.	• More blood flows near the skin surface and gives off heat to the environment (**vasodilation**). • More sweating. The water in sweat **evaporates**. This takes heat from the skin and transfers it to the environment. • Wearing fewer clothes allows heat loss.

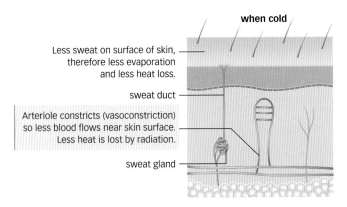

when cold

Less sweat on surface of skin, therefore less evaporation and less heat loss.

sweat duct

Arteriole constricts (vasoconstriction) so less blood flows near skin surface. Less heat is lost by radiation.

sweat gland

▲ When the body needs to gain heat, you sweat less and vasoconstriction brings less blood near the skin surface

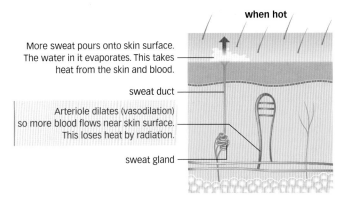

when hot

More sweat pours onto skin surface. The water in it evaporates. This takes heat from the skin and blood.

sweat duct

Arteriole dilates (vasodilation) so more blood flows near skin surface. This loses heat by radiation.

sweat gland

▲ When the body needs to lose heat, you sweat more and vasodilation brings more blood near the skin surface

Negative feedback

Negative feedback is a mechanism that keeps things at a steady level. It works like this:

• The level changes away from steady.
• Sense organs in your body detect the change.
• They send information to a control centre in the brain.
• This sends information back to particular body structures.
• That sets processes in motion that redress the balance and bring the level back to steady.

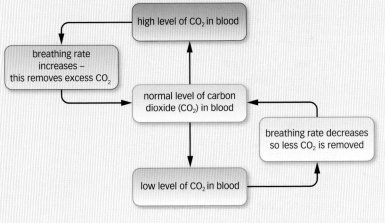

high level of CO_2 in blood

breathing rate increases – this removes excess CO_2

normal level of carbon dioxide (CO_2) in blood

breathing rate decreases so less CO_2 is removed

low level of CO_2 in blood

▲ Negative feedback keeps a steady level of carbon dioxide in your blood

Exam tip **OCR**

✔ Remember it is the *water* in sweat that evaporates.

Questions

1 List five factors that need to be kept at steady levels in the body.
2 What is the normal core body temperature for humans?
3 State three ways the body loses heat and three ways it gains heat.
4 What is homeostasis?
5 Explain how vasodilation helps to keep you cool.

E

A*

Learning objectives

After studying this topic, you should be able to:

- ✔ explain why the body's reactions to hormones are usually slower than nervous reactions
- ✔ state that insulin is made in the pancreas
- ✔ explain how insulin helps to regulate blood glucose levels
- ✔ recall how diabetes is caused and how it can be controlled

Key words

hormone, gland, target organ

A What are hormones?

B Why does the body usually respond slowly to hormones?

Hormones

Hormones are chemicals. They

- are secreted from **glands** into the bloodstream
- travel in the blood to **target organs**
- regulate the functions of many organs and cells
- coordinate many processes in the body.

The body usually reacts more slowly to hormones than to nerve impulses, as hormones coordinate long-term changes such as maturing (growing up). Hormones have to be synthesised and released from the cells in the body where they are made, and then travel all over the body, in blood, to reach their target organs. Nerve impulses are generated very quickly and travel via neurones directly to the effector.

However, some hormones act quickly. Adrenaline, which your body makes when you are frightened, acts quickly.

You can see from this diagram that the pancreas makes a hormone, insulin. Insulin controls the amount of glucose (sugar) in the blood. The blood must have the right amount of glucose so it can deliver it to the cells. The cells can then get energy from the glucose.

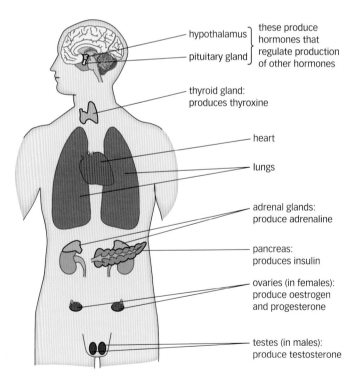

hypothalamus ⎫
pituitary gland ⎭ these produce hormones that regulate production of other hormones

thyroid gland: produces thyroxine

heart

lungs

adrenal glands: produce adrenaline

pancreas: produces insulin

ovaries (in females): produce oestrogen and progesterone

testes (in males): produce testosterone

▲ Body outline showing position of the main organs and some endocrine glands, and the hormone each produces

Insulin

Insulin regulates the blood glucose level.

> If there is too much glucose in the blood:
> - the pancreas secretes insulin
> - which travels in the blood to the liver
> - and makes the liver take up the extra glucose and change it to another carbohydrate called glycogen. This is a way of storing energy.

If your pancreas does not make enough insulin this causes a disease called diabetes.

Diabetes

There are two types:

Type of diabetes	Cause	Method of control
Type 1	The person's own immune system has destroyed the cells in the pancreas that make insulin.	Controlled with insulin injections at mealtimes. Patients should also eat a balanced diet and take exercise.
Type 2	Occurs in later life. It may be linked with being overweight. The person's cells do not respond to insulin and their pancreas may also not make enough insulin.	May be controlled by eating a balanced diet with more fibre and less sugar. People may also take medicines to make their pancreas make more insulin. In some cases they may need insulin injections.

People with diabetes have to test their blood to see how much glucose is in it. They can inject themselves with insulin after meals if it rises too high. They can also make sure that they do not eat too much sugary food.

> Diabetes sufferers can also control their symptoms by exercising. Some excess glucose in the body is converted to energy during exercise. The amount of insulin a person needs to inject depends on how much sugar they eat and how active they are.

Exam tip

- ✓ If you are asked what a hormone is, give as much information as possible. Don't just say it is a chemical. Say they are made in glands and that they travel in the blood to target organs, to coordinate body processes.

Questions

1. Which part of the body produces insulin? ↓ E
2. (a) What causes type 1 diabetes?
   (b) How is it treated?
3. (a) What causes type 2 diabetes? ↓ C
   (b) How is it treated?
4. Describe how insulin lowers the amount of glucose in the blood.
5. Insulin used to be obtained from dead pigs' pancreases. Nowadays it is made by genetically modified bacteria. What do you think are the advantages of making it in this way? ↓ A*
6. Explain why someone with diabetes who takes a lot of exercise would need less insulin than someone who does not take exercise.

Learning objectives

After studying this topic, you should be able to:

- ✔ know that plants are sensitive to light, moisture, and the force of gravity
- ✔ understand that plants produce hormones to control and coordinate growth
- ✔ recall how plant hormones can be used in agriculture

▲ Seedlings growing towards light. They are showing a positive phototropic response.

A   What do plant growth hormones control in plants?

B   What is (a) phototropism (b) geotropism?

Key words

tropism, phototropic, auxin, geotropic

Plants respond to their environment

Plants as well as animals respond to stimuli, changes in their environment. Plants make chemicals called plant hormones (plant growth substances) that control and coordinate

- the growth of shoots and roots
- flowering
- ripening of fruits.

You may have noticed that plants growing on a windowsill tend to bend towards the light source. If you want them to grow straight you have to keep turning them. A plant's response to a stimulus is called a **tropism**.

Phototropism

Plant shoots grow towards light. They are positively **phototropic**.

- They make a plant hormone called **auxin**

  - This moves through the plant in solution.
  - When light strikes one side of the shoot tip, more auxin builds up on the *other* side of the shoot tip (furthest from the light).
  - This causes the shoot to bend over, towards the light.
  - This increases the plant's chance of survival because it needs light to photosynthesise and make food.

How does auxin make the shoot bend?

Auxin is made in the shoot tip and is unevenly distributed when the shoot is exposed to light from one side. This auxin moves down the stem and causes cells on the side of the shoot furthest from the light to elongate more than those nearest the light.

Geotropism

Geotropism is the response of a plant to gravity. Roots grow downwards in response to the pull of gravity. Auxins may be involved in this response but other chemicals that inhibit growth may also play a part.

Plant hormones in agriculture and horticulture

Plant hormones can be applied to plants to either speed up or slow down their growth.

▲ Venus flytrap

▲ The sensitive plant's leaves close up after the plant has been touched

Did you know...?

Plants also respond to other stimuli. Plant roots grow towards water.

Some plants respond to touch. Climbing plants put out tendrils and cling to walls or canes.

Some plants move their leaves downwards very quickly when touched. This startles animals trying to eat them and protects the plant. The Venus flytrap responds to flies touching hairs on its leaves. It snaps the leaves shut, trapping the fly so it can digest it.

Weedkillers

Auxins are used as selective weedkillers. Agent Orange was used in the Vietnam War. It made trees lose their leaves. Without leaves, plants cannot make food and the trees die.

Rooting powder

If cuttings are dipped into auxin powder, this helps the cuttings make new roots. With new roots the cuttings can anchor in the soil and take up water and minerals.

Fruit ripening

Auxin can be sprayed onto fruit trees to prevent the ripe fruit from dropping. Then it can all be harvested at the same time and the fruit is not bruised by falling to the ground. If a high dose of auxin is sprayed later, it makes all the fruit fall.

Control of dormancy

Seeds are usually dormant when they are shed from the parent plant. This stops them germinating at the end of the summer, as new plants may not survive the harsh winter. However, if commercial growers want to germinate seeds in greenhouses during winter, applying auxin can break the dormancy.

Exam tip **OCR**

✓ Remember that although plant growth substances like auxin are called plant hormones, they do not work in the same way as animal hormones.

Questions

1   Why is it an advantage to a plant if the shoots are positively phototropic?

2   Why is it an advantage to a plant if the roots are positively geotropic?

3   Describe four commercial uses of auxins.

4   Explain how auxin causes shoot tips to bend towards the direction of the light source.

▲ Horticulturist dipping a cutting in rooting powder

Learning objectives

After studying this topic, you should be able to:

- ✔ know that differences in characteristics may be due to differences in genes or the environment or both
- ✔ recall that body cells have the same number of chromosomes but this number varies between species

Key words

gene, chromosome, gamete

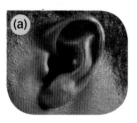 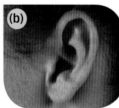

(a) (b)

▲ The shape of your earlobe is determined by genes. (a) attached earlobe (b) free hanging earlobe.

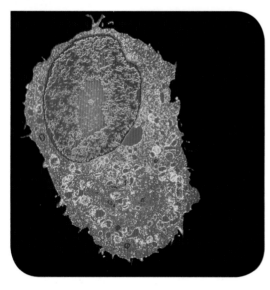

▲ A mammalian cell (×400). The cytoplasm is coloured blue, and the large orange and green structure is the nucleus. Inside the nucleus are chromosomes containing genes.

How human characteristics are determined

Some characteristics are determined by the environment, some by genes, and some by both genes and environment.

Characteristics determined by the environment

These include

- scars
- learning to speak a language.

You get scars after you have injured yourself. Your children will not be born with the same scars.

At birth you cannot speak: you have to learn by listening to your parents and copying them. If you never hear anyone speak you will not be able to master language.

Characteristics determined by genes

These characteristics may be inherited. They include

- eye colour
- earlobe shape
- nose shape.

Characteristics determined by both genes and environment

Some characteristics depend on both genes and environmental factors. For example, you may inherit genes that control your brain development in such a way that you will be able to have high intelligence. However, if you are not fed properly whilst growing up, or if no one talks to you or reads to you, and you are not given opportunities for stimulating play, you will not develop your full intelligence potential. Height and body mass are also determined by both genes and environment. You may have the genetic potential to be very tall but you will not fulfil it if you are undernourished.

Genes and chromosomes

Genes are particular regions of **chromosomes**, the thread-like structures inside the nucleus of each cell. Chromosomes are made of a chemical called DNA. Each gene is a length of DNA. Each gene controls the development of a particular characteristic.

All your body cells have the same number of chromosomes (23 matched pairs) in the nucleus, except your red blood cells, which do not have a nucleus.

The number of chromosomes in each cell varies between species. Dogs have 39 matched pairs, horses have 32, and chimpanzees have 24.

Gametes, eggs, and sperms, have half the number of chromosomes of body cells.

The nature–nurture debate

Many people have debated for a long time whether genes (nature) or environment (nurture) are more important in determining certain characteristics, such as intelligence, sporting ability, health, and criminality. Most scientists now think that both these factors work together in determining these characteristics. With all the information coming from genetic research, for example the Human Genome Project, it seems that genes play a very important role in many aspects of our behaviour – even the way we think. This may be why it is hard to change people's attitudes to things like racism or adopting a healthy lifestyle.

Questions

1 State three human characteristics that are determined by (a) the environment (b) genes (c) both genes and environment.
2 What are chromosomes?
3 What are genes?
4 How many chromosomes do you have in each of your (a) liver cells (b) brain cells (c) white blood cells?
5 What is 'the nature–nurture debate'?

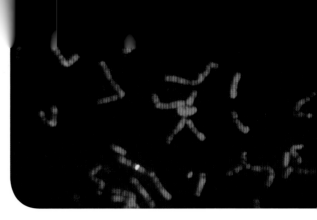

▲ Chromosomes. Some of the genes on some chromosomes have been tagged with a fluorescent chemical and show up as yellow on the red chromosomes.

▲ Echidna (spiny anteaters, egg-laying mammals related to the duck-billed platypus) are unusual: females have 64 chromosomes in each cell nucleus, and males have 63.

Exam tip OCR

✓ Remember to use information given to you in a question. You may for example be given information about chromosomes in the cells of an animal of another species. Do not just recall information about human chromosome number, but mention this species as well.

Learning objectives

After studying this topic, you should be able to:

- ✔ explain that genetic variation can be caused by mutations, gamete formation, and fertilisation
- ✔ recall that gametes have half the number of chromosomes of body cells

Key words

mutation, fertilisation, spontaneous, allele, sexual reproduction, random

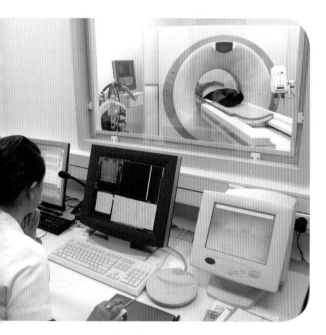

▲ CT scans provide very useful diagnostic tools. However, the radiographer goes out of the room so that they will not be exposed to too many X-rays while carrying out their work.

What causes genetic variation?

Mutations, formation of gametes, and **fertilisation** can all produce variation.

Mutation changes genes

You can see from the picture on spread B1.13 that some people have free earlobes and some have attached lobes. A gene determines the characteristic earlobe shape. Like all genes, this one is a length of DNA on a particular chromosome. Sometimes a mutation changes the structure of a gene.

Some mutations just happen **spontaneously** – there is no external cause. A mistake happens when DNA is being copied before cells divide.

Some mutations are caused by things in the environment such as

- chemicals (tar) in tobacco smoke
- chemicals used for dying materials
- ultraviolet radiation in sunlight
- ionising radiation like X-rays and gamma rays.

A What is a mutation?

B Name three things that can cause mutations.

Alleles

A mutation may change a gene to a different version, called an **allele**. It still codes for the same characteristic but will produce a slightly different version of the characteristic.

The gene for earlobe shape has two alleles: one codes for free lobes and the other codes for attached lobes.

Sexual reproduction also causes genetic variation

Sexual reproduction involves fertilisation, the joining of gametes.

Gametes

Gametes are special sex cells. They contain only 23 chromosomes. This is half the number of chromosomes in normal body cells.

- Female gametes are eggs.
- Male gametes are sperms.

Gametes are made from body cells by a special type of cell division. The body cell has 23 pairs of chromosomes. The pairs separate during the cell division. Gamete formation produces genetic variation because

- during the cell division, the chromosomes in each pair may swap pieces with each other, giving them different combinations of alleles, but still retaining the same genes
- the way the two copies (chromatids) of each chromosome separate from each other during the second meiotic division is random.

This means that every egg a woman produces is genetically different, and every sperm a man produces is genetically different. Each gamete is genetically unique.

Fertilisation

A male and female gamete join by fertilisation to make a new individual. Fertilisation is **random** and produces genetic variation. Each egg produced by the mother is genetically different from the other eggs she makes. The father makes many sperms and any one of them could join with an egg. As a result every individual will be genetically unique.

One male gamete (sperm) and one female gamete (egg) fuse. The resulting fertilised egg, which will become a new individual, contains the full number of chromosomes. Half have come from the mother and half from the father. So the mixture of genetic information from two parents leads to genetic variation in the offspring.

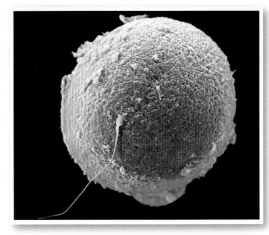

▲ Fertilisation. The sperm and the egg fuse to create a new genetically unique individual. (× 700).

Exam tip — OCR

✔ If explaining that random fertilisation produces unique individuals, describe them as *genetically* unique. Similarly, when talking about the variation amongst organisms produced by sexual reproduction, say they have *genetic* variation.

Questions

1 What are gametes?
2 What happens to the number of chromosomes at fertilisation?
3 Explain why the radiographer goes out of the room while the patient is being scanned.
4 Explain why fertilisation is random.
5 What are alleles?
6 Explain why sexual reproduction produces genetically unique new individuals.

Key words

dominant, recessive, **genotype**, **phenotype**, **heterozygous**, **homozygous**, **monohybrid cross**

A A dog breeder crossed a dog with long hair with one having short hair, several times. All the puppies had long hair. Which characteristic is dominant?

B If someone has sex chromosomes XY, are they male or female?

	Sex chromosomes in male gamete	
	Ⓧ	Ⓨ
Sex chromosomes in female gamete		
Ⓧ	XX 50% chance of female offspring	XY 50% chance of male offspring

Dominant and recessive inherited characteristics

Biologists often carry out breeding experiments to see how characteristics are inherited.

Imagine you carried out several experiments crossing pure-breeding white mice with pure-breeding brown mice. If all the offspring were brown, you would know that brown coat is the **dominant** characteristic and white coat is the **recessive** characteristic.

Dominant and recessive alleles

The offspring mice all have an allele for brown coat and an allele for white coat, but the brown coat is expressed (seen). We can say that the allele for brown coat is dominant and the allele for white coat is recessive.

What makes us male or female?

You know that you have 23 pairs of chromosomes in the nucleus of all your body cells. One of these pairs of chromosomes determines sex (gender). If, in that pair, you have two large X chromosomes, you are female. If you have one large X chromosome and a smaller Y chromosome, you are male.

Inheritance of sex

We can use a genetic diagram to show how sex (gender) can be inherited.

Parents	Male	Female
Parents' sex chromosomes	XY	XX
Sex chromosomes in gametes	Ⓧor Ⓨ	allⓍ
Possible combinations at fertilisation	XX or XY (SEE LEFT)	

You can see that half the male gametes (sperms) have an X chromosome and half have a Y chromosome. However, all the female gametes (eggs) have an X chromosome. Fertilisation is random, and either type of sperm could fertilise an egg. At each pregnancy there is a 50:50 chance of conceiving a girl or a boy. In a large population, there will be equal numbers of male and female offspring.

Homozygous and heterozygous

The genes you inherit from your parents determine your **genotype**. The physical expression of the genes you carry for a particular characteristic (such as earlobe shape) is your **phenotype**.

For example, you might inherit one chromosome from your father that has a gene coding for free earlobes, and a different allele (version of the gene) from your mother, coding for attached earlobes. You have two different versions of the gene for earlobe shape (you are **heterozygous** for this characteristic), but you will have free earlobes because this characteristic is dominant. The allele for free lobes will be expressed. Attached lobes are a recessive characteristic. You will only have attached earlobes if you carry two alleles for attached lobes (ie you are **homozygous** for this characteristic).

This type of cross where we just look at the inheritance of one characteristic is called a **monohybrid cross**.

Genetic diagram for a monohybrid cross

There are conventions for drawing genetic diagrams:
- Show the characteristic of the parents.
- Show the alleles present in the parents' cells.
- Use upper case letters to represent a dominant allele.
- Use lower case version of the same letter to show a recessive allele.
- Put gametes in circles.
- Show all the different possible combinations of alleles at fertilisation.
- Put an 'x' to show a cross (mating).

Here is a genetic diagram to explain a monohybrid cross between a father who is homozygous for free earlobes and a mother who is homozygous for attached earlobes.

Parents' characteristics	father × mother free earlobes × attached earlobes
Parents' alleles	**EE** × **ee**
Gametes	all (**E**) all (**e**)
Offspring's alleles	all **Ee**
Offspring's characteristics	all have free earlobes (all are heterozygous)

Exam tip   OCR

✔ Always use the conventions when drawing genetic diagrams. Also make sure your upper case letters are big and lower case letters are small. If you are told to use particular letters then make sure you use them.

Questions

1 Explain how sex is determined in humans. ↓ **C**

2 A couple have three children – all girls. What are the chances of them having a boy at the next pregnancy?

3 A mother with blue eyes and a father with brown eyes have three children. Two have brown eyes and one has blue eyes. Brown eyes is the dominant characteristic. Draw a genetic diagram to explain how these parents can produce these offspring. ↓ **A***

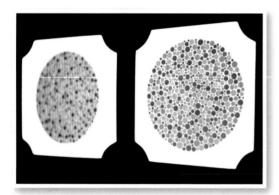

Cards used to test for red-green colour blindness. People with red-green colour blindness cannot see the numbers 29 and 57.

Did you know...?

There are about 5000 genetic disorders caused by faulty alleles, but the incidence of each in the human population is rare. These are all harmful, although some are more harmful than others.

Inherited disorders

Inherited disorders are caused by faulty genes.

Examples of inherited disorders include

- cystic fibrosis
- red-green colour blindness.
- sickle cell anaemia

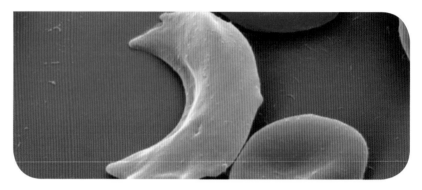

▲ Electron micrograph of sickled red blood cell next to some normal red blood cells (× 10 000)

Faulty alleles

Most (but not all) of the faulty alleles that cause inherited disorders are recessive, so a person only has the disorder if they inherit two faulty alleles. It may be that neither parent had the disease, but both had a faulty allele and were carriers.

> **A** Name three inherited disorders in humans.
>
> **B** What causes inherited disorders?

Not all mutations are harmful

Some mutations may be beneficial (useful). Here are some examples:

- Early humans lived in Africa and had dark skin which protected them from sunburn. A mutation that gave them pale skin would be harmful. However, as humans migrated to areas with less intense sunlight, pale skin was useful. This is because it would allow the sun's less intense rays to penetrate the skin enough to make vitamin D. Vitamin D allows us to absorb calcium from food. This prevents rickets.
- A mutation to a gene known as the FOXP2 gene enabled humans to talk and develop speech.

Will inherited disorders pass to the next generation?

We can use genetic diagrams to predict the probabilities of inherited disorders passing to the next generation.

Suppose both members of a couple have family members with cystic fibrosis. They could both have a test to see if they carry a faulty allele for cystic fibrosis. If both are carriers of the faulty allele, they can work out the probability of their baby having cystic fibrosis.

Both are carriers so they have the alleles: **Cc** × **Cc**

Their gametes are: Ⓒ Ⓒ × Ⓒ Ⓒ

A Punnett square shows what could happen:

Gametes	C	c
C	CC normal	Cc carrier
c	Cc carrier	cc cystic fibrosis

So, at each pregnancy there is a 1 in 4 or 25% chance that the baby will have cystic fibrosis.

Issues raised by knowledge of inherited disorders in the family

Everyone will have some mutations in their chromosomes but most are for a very rare disorder and the chance of having a child with someone else who has the same mutation is extremely small. However, if people know they have an inherited disorder in the family they have to think about whether they should

- tell their partner
- both have a genetic test to find out if they are carriers of the disorder
- have the fetus tested and consider terminating the pregnancy.

What they decide may depend on their personal choice or religious views. A genetic counsellor will be able to give them information so that the couple can make a more informed choice.

Questions

1 Give an example of a beneficial mutation. ↓ E

2 A couple is considering having a baby. Discuss the issues raised by a couple knowing that they each have a severe inherited disorder in the family. ↓ C

3 One parent carries a faulty allele for sickle cell anaemia. The other parent does not have this allele. Draw a genetic diagram to predict the chances of their having a child with sickle cell anaemia.

4 Explain how faulty alleles can change a characteristic.

5 In a couple, both have an allele for sickle cell anaemia but neither suffers from the disease. Draw a genetic diagram to predict the chances of their having a child with sickle cell anaemia. ↓ A*

6 In the UK it is legal for first cousins to marry. First cousins share grandparents. Explain why the incidence of inherited disorders may be higher amongst first-cousin marriages.

Module summary

Revision checklist

- Heart disease is a major killer in the UK and is associated with high blood pressure.
- A balanced diet contains the right amount of energy and enough of each nutrient.
- Infectious diseases are caused by microbes. Disease can also be caused by vitamin and mineral deficiencies, body disorders like cancer, and inherited genetic disorders.
- Some fungi, bacteria, viruses, and protozoa cause diseases. They are more common in hot, wet countries with poor sanitation.
- Our bodies keep most pathogens out. White blood cells ingest pathogens and produce antibodies to destroy them. Vaccines provide immunity to specific infections.
- Our eyes detect light and our brains form an image. Having forward-facing eyes allows binocular vision, which makes distances easier to judge.
- Our brains control most actions but we also have some automatic reflex actions, which protect us.
- All drugs alter your body or brain. They can have harmful side effects or be addictive.
- Alcohol is a depressant and can cause liver and heart disease or brain damage. Cigarettes are addictive and cause lung damage.
- We keep our temperature, water content, and carbon dioxide levels steady to keep conditions right for the chemical reactions in cells. This is homeostasis.
- Hormones regulate the functions of many organs and cells. They include insulin, which regulates blood sugar.
- Plants also use hormones to respond to change.
- Our characteristics are controlled by inherited genes and environmental factors, like our nutrient intakes.
- Sexual reproduction combines genes from each parent, generating genetic variation. Mutations also cause variation.
- Genes come in pairs. If they are different, one may be dominant, which means that it controls the characteristic that the gene codes for.
- Inherited disorders like cystic fibrosis are caused by faulty genes. Some mutations make genes more useful.

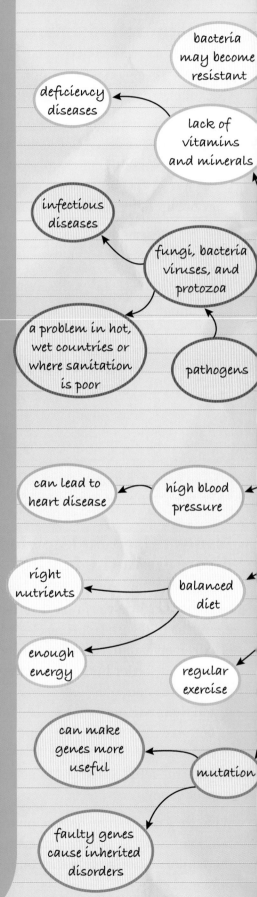

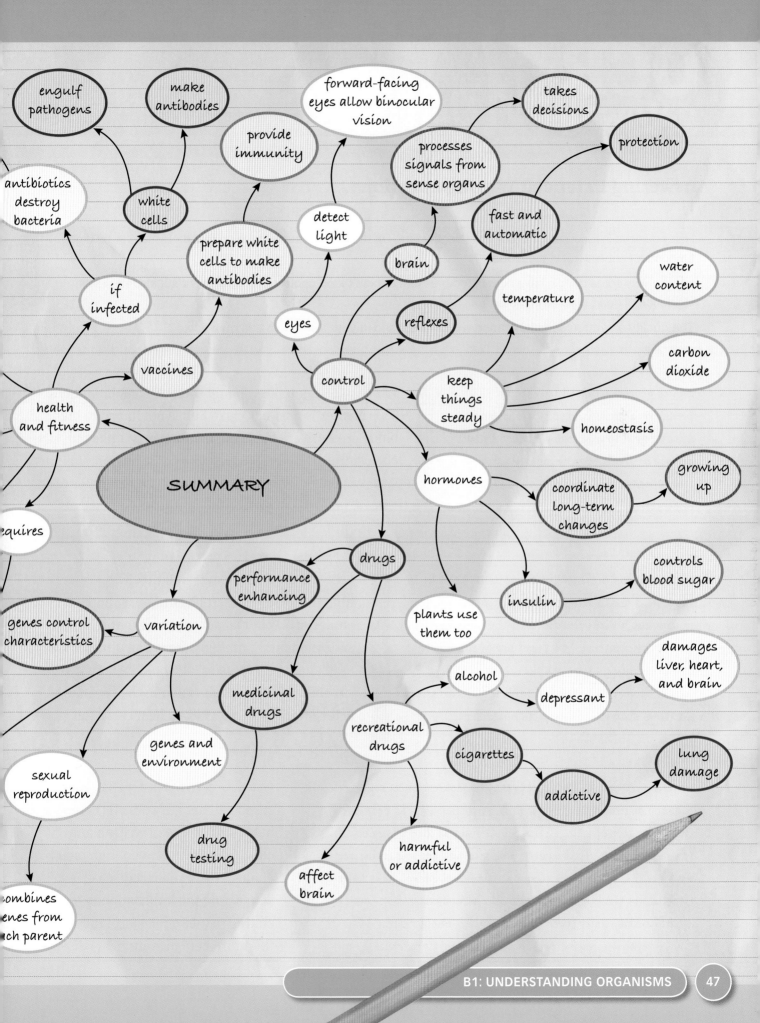

engulf pathogens

make antibodies

forward-facing eyes allow binocular vision

takes decisions

provide immunity

processes signals from sense organs

protection

antibiotics destroy bacteria

white cells

detect light

fast and automatic

prepare white cells to make antibodies

brain

water content

if infected

eyes

reflexes

temperature

carbon dioxide

vaccines

control

keep things steady

homeostasis

health and fitness

SUMMARY

hormones

coordinate long-term changes

growing up

requires

drugs

insulin

controls blood sugar

performance enhancing

genes control characteristics

variation

plants use them too

damages liver, heart, and brain

medicinal drugs

alcohol

depressant

genes and environment

recreational drugs

cigarettes

lung damage

sexual reproduction

drug testing

affect brain

harmful or addictive

addictive

combines genes from each parent

Answering Extended Writing questions

QUESTION

Heart disease is the main cause of early death in the UK. It may be caused by high blood pressure or by high amounts of cholesterol in the blood.

Explain why some people have high blood pressure or high blood cholesterol. How might statins reduce the incidence of heart disease in the UK?

The quality of written communication will be assessed in your answer to this question.

Some people inherit high blood pressure. you can also get it if you smoke and drink. eating a lot of salt or lots of foods like chips can give you a heart attack. statins lower the cholesterol in your blood.

↓ E

Examiner: Sentences should start with capital letters. This answer is not very clear. Salt is not linked to high blood pressure and fat is not mentioned or linked to cholesterol. 'Drinking' is also vague; the answer should specify alcohol. This answer requires the examiner to do a lot of work and interpret what the candidate means.

If your blood pressure is above 140/95 it is high and could cause a heart attack.
If you eat a high fat diet then your liver makes cholesterol. You need some cholesterol to make things like sex hormones. If there is lots of it in your blood it can clog up arteries. Statins stop you making cholesterol.

↓ C

Examiner: This hasn't explained what causes high blood pressure. It has linked fat and cholesterol, but has not specified that it is saturated fat. Examiners don't like phrases such as 'clogging up arteries'. It would be better to say that if fatty deposits form in the coronary artery walls, you may have a heart attack.

People may get high blood pressure because they smoke tobacco or drink lots of alcohol. Another way is because they eat a lot of salt. Some people have it due to genetics. If you eat a lot of saturated fat, such as red meat, butter and cream, your liver turns it into cholesterol. It is taken to cells, in the blood, to make cell membranes stronger. Statins stop your liver from making cholesterol. Cholesterol in your blood can cause heart attacks.

↓ A*

Examiner: This is a well organised response that answers all three questions. It is well written and clear. The link between high blood cholesterol and heart disease is stated, and the implication is that if statins reduce cholesterol they will therefore help prevent heart disease.

Exam-style questions

1 The diagrams show three organisms that can cause disease.

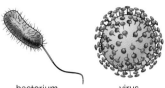

bacterium   virus   protozoan

A01 **a** Which causes flu?

A01 **b** Which is destroyed by antibiotics?

A01 **c** Which causes cholera?

A01 **d** Which could be spread by mosquitoes?

A01 **e** Name the type of microbe that causes athlete's foot.

A01 **f** List three ways microbes are kept out of our bodies.

2 Scientists asked twenty 13–19-year-olds about the sports they played and whether they had ever smoked. Then they tested their lungs. The smokers had smaller lung volumes and more of them had quit sports.

A01 **a** List four harmful components of cigarette smoke.

A01 **b** Which substance limits a smoker's oxygen supply?

A01 **c** Which two substances could be damaging the smokers' lungs?

A02 **d** Why was it important to compare smokers and non-smokers of similar ages?

A02 **e** How could you change the test procedure to make the results more reliable?

A03 **f** Why might smokers find it hard to give up even when they notice smoking is harming their health?

3 Many children inherit genetic disorders that their parents don't have. Their parents are carriers, and the disorders are caused by recessive alleles.

A01 **a** Explain what a recessive allele is.

A01 **b** What name is given to people who carry one recessive allele and one dominant one?

A02 **c** Use C to represent the dominant allele and c to represent the recessive allele. What genotype would a carrier have?

A02 **d** Two carriers are planning to have a family. Use a genetic diagram to predict the probability of their having a child with the disorder.

Extended Writing

4 Dev cut his hand in the garden. His
A01 finger is red and swollen. It's infected. Describe how his body cells will deal with the infection. Use as many scientific words as possible.

5 Kevin is taking part in a drug trial.
A02 Scientists want to know whether a new antibiotic (N) is better than an older one (O), for treating severe throat infections. Twenty people with throat infections will stay at the test centre for a week. Write a set of instructions for the nurses running the trial.

6 Many people cannot make enough of a
A01 hormone called insulin. Explain what role the hormone plays in the body and what can be done to help people with this disorder lead normal lives.

B2

Understanding our environment

Why study this module?

There are millions of different organisms on this planet. They all interact with each other and with their environment. The best adapted organisms are the ones most likely to survive.

In this module you will learn about the amazing range of living organisms, and how biologists are able to sort this variety of life into groups. Along with grouping, you will study the ideas of feeding relationships and energy flow through the food chain. Death and decay are part of this, and so is recycling.

The ways in which different organisms compete with each other for limited resources will be investigated. You will learn about the fantastic ways that plants and animals have become adapted to survive. You'll then learn about how all this life evolved. Finally, you will examine some of the big issues of the day: the effects of human pollution, and how we can all live a sustainable lifestyle.

You should remember

1 Plants and animals are sorted into groups based on their characteristics. This is called classification.

2 Feeding relationships are shown in food chains and webs.

3 Energy is passed through the food chain by animals feeding.

4 Organisms depend on each other for their survival.

5 Plants and animals are adapted to survive in their environment.

Every organism is superbly adapted to life in its environment. An extreme example of ultra efficient design is the cheetah.
It can accelerate from 0 to 100 km/hour in 3.5 seconds, faster than a Ferrari. Their top speed of 110 kph (70 mph) is the fastest of any land animal. But the cheetah still has to compete with others for its food. Unfortunately, after it runs it is often too tired to defend its kill.

A Humans are animals.
   (a) Give three reasons why we are classified as animals.
   (b) Which subgroup of animals do humans belong to? Explain why.

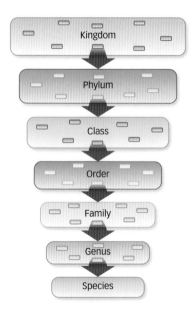

Dividing the kingdoms. A kingdom is divided into many phyla, a phylum into many classes, a class into many orders, and so on.

Similarities and differences

The world is full of millions of different types of living things. Biologists put living things into groups, which makes them easier to study. This grouping process is called **classification**.

To classify living things biologists observe their features (characteristics). You can observe both similarities and differences in these two types of daffodil.

◄ Similarities:
 • same shaped leaves
 • same number of of petals
Differences:
 • different colour
 • different height

Living things that share lots of similar characteristics are grouped together. If organisms have lots of differences in their characteristics then they are classified in different groups.

The kingdoms

All living things are placed into five major groups called **kingdoms**.

Kingdom	Characteristics	Example organisms
plants	– made of many cells – cell walls made of cellulose – cells contain chloroplasts – use light to make food by photosynthesis – grow in a spreading manner	Flowers, mosses, grasses
animals	– made of many cells – no chloroplasts – unable to make food so feed on other organisms – compact body shape – most are capable of moving around	Fish, insects, humans
fungi	– cell walls made of chitin – reproduce by forming spores – do not photosynthesise	Mushrooms, toadstools, yeast
protoctista	– mostly single-celled (except the large algae)	Amoeba, algae
prokaryotes	– no nucleus	Bacteria

Animal kingdom

There are just over a million different types of animal in the world.

Biologists decided on a way of dividing the animal kingdom into many phyla, based on animals' physical characteristics (eg all **vertebrates** have a backbone). Phyla include:

- cnidarians – eg jellyfish
- annelids – eg earthworms
- arthropods – eg insects
- molluscs – eg slugs and snails
- vertebrates – eg fish, frogs, dinosaurs, and humans.

A closer look at arthropods

The **arthropods** are the largest animal phyla. They represent about 75% of animal species and are divided into four classes.

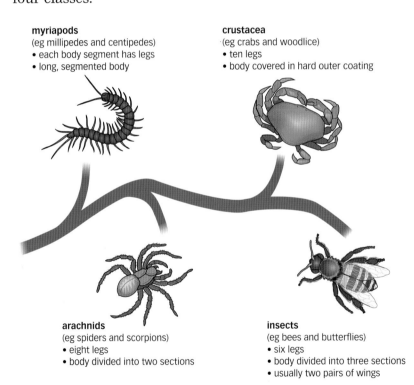

myriapods
(eg millipedes and centipedes)
- each body segment has legs
- long, segmented body

crustacea
(eg crabs and woodlice)
- ten legs
- body covered in hard outer coating

arachnids
(eg spiders and scorpions)
- eight legs
- body divided into two sections

insects
(eg bees and butterflies)
- six legs
- body divided into three sections
- usually two pairs of wings

Problems with classifying

Classification is not always as simple as it sounds. There are organisms that do not fit neatly into one group. They may share characteristics from two groups. You may have heard of an animal called *Archaeopteryx*. Biologists believe it is the link between reptiles and birds. There is a large range of organisms, and many organisms are quite similar.

Classification systems

Biologists use two systems in classifying organisms:

- Natural – based on evolutionary links, so organisms are grouped together if they have common ancestry.
- Artificial – based on visible features, making a group of organisms easier to identify. All yellow flowering plants could be grouped together, for example.

As biologists learn more about organisms, such as their genetics and the sequence of their DNA bases, we realise that some organisms are more closely related than originally thought. A natural classification needs to adjust to form more accurate links between organisms.

Questions

1 Why do biologists classify living things?

2 Draw a simple diagram of a bee. Label the characteristics that put it in the arthropod class of insects. ↓ E

3 Every day new organisms are discovered. How would you go about deciding in which group a new discovery should be placed ? ↓ C

4 What is the difference between a natural and an artificial classification system?

5 Explain how advances in genetics have helped with the process of classification. ↓ A*

Common name	Genus	Species
human	*Homo*	*sapiens*
tiger	*Panthera*	*tigris*
lion	*Panthera*	*leo*
buttercup	*Ranunculus*	*repens*

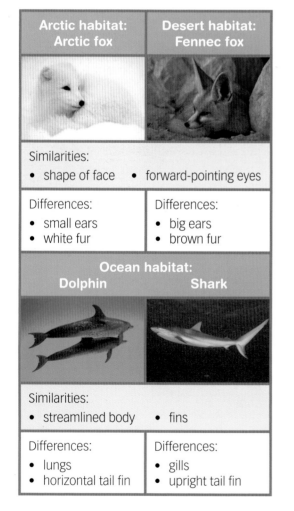

Arctic habitat: Arctic fox	Desert habitat: Fennec fox

Similarities:
• shape of face • forward-pointing eyes

Differences:	Differences:
• small ears	• big ears
• white fur	• brown fur

Ocean habitat:	
Dolphin	Shark

Similarities:
• streamlined body • fins

Differences:	Differences:
• lungs	• gills
• horizontal tail fin	• upright tail fin

What is a species?

As humans we all belong to one species. Dogs are very different from humans and they belong to a different species. But what is a species?

• A **species** is a group of similar organisms that are capable of interbreeding to produce fertile offspring.

So members of the dog species can interbreed with each other to produce more dogs, which can also interbreed. Dogs and cats are different species; they cannot breed together.

The binomial system of naming species

Biologists have a way of naming all organisms, called the binomial system. Each species is known by a two-part name. Scientists all over the world use the same name for a species, which prevents confusion. The first name, the genus, is a group name for closely related species. The second name identifies a single species.

Variation within a species

Individuals within a species are not all the same. In the dog species there is a huge variety of different dogs. Dogs have been bred for their looks or their ability to work. But although dogs show **variation**, they have more features in common with each other than they share with other species. This is an evolutionary link because they share a common ancestor.

Variations to suit a habitat

Often, similar species live in similar habitats. However, closely-related species may look very different if they live in different habitats. The Arctic fox lives in an Arctic habitat and the fennec fox lives in a desert habitat. They still share certain characteristics and are classified close together.

On the other hand, organisms that are not closely related may share several features if they live in the same habitat. For example, the shark and the dolphin both have an ocean habitat. These show an ecological link because the two unrelated species are not classified together, but need to solve the same environmental problems.

Evolutionary links between species

If two species are closely related, then they tend to share more characteristics. This is because they both had the same close ancestor. Chimpanzees and humans share a number of features. They both

- have a backbone
- feed their young with milk
- have skin covered in hair
- have grasping hands.

This is because they both had the same common ancestor. If you compare these two species with a lizard, there are far fewer similarities. They all have a backbone, for example. The more characteristics that are analysed, the more accurate the family tree model becomes. Computers now analyse large numbers of characteristics for us.

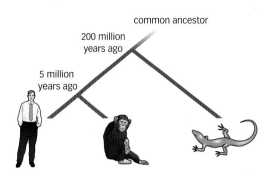

▲ This family tree shows the relationship between three species. The human and chimp are far more closely related to each other than they are to the lizard.

Problems with classifying species

Hybrids

Sometimes individuals from two closely related species can breed together. The horse can breed with the donkey to produce a mule. The offspring is a mixture of the features of the two different parents. This type of offspring is called a **hybrid**. Often, hybrids are unable to breed themselves.

Classifying asexual organisms

Bacteria are an unusual case. There are many species of bacteria, and they do not reproduce with one another. They reproduce asexually. Biologists still classify them as different species, based on their characteristics.

Constant evolution

Evolution means that organisms are slowly changing. They can eventually become new species.

▲ Mule

Questions

1. Name the system biologists use to name organisms.
2. There are lots of varieties of horse. Explain how biologists know that they are the same species.
3. Donkeys and horses look very similar. Suggest why this is.
4. Explain how biologists know that donkeys and horses are different species.

A What is a species?

B What is the advantage of having a universal binomial system?

Key words

species, variation, **hybrid**

A Construct food chains for the following organisms:
hawk, leaf, bluetit, caterpillar;
seaweed, crab, seagull, periwinkle;
fish, pondweed, heron.

B Name the producers, primary consumers, secondary consumers, and top carnivores in each of the food chains from Question A.

C Explain why a food web is more useful to a biologist than a food chain.

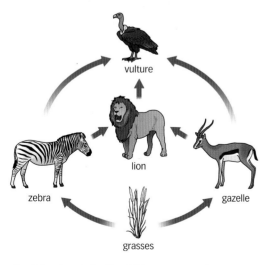

▲ A food web for the African savannah

Food chains

A **food chain** shows what organisms eat – it shows the flow of food and energy from one organism to the next. The arrows show the direction of flow of food and energy. Each link in the food chain is a separate feeding level called a **trophic level**.

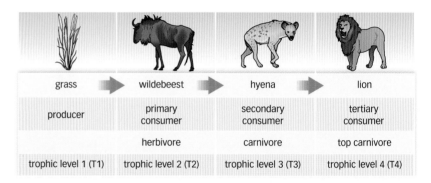

grass	wildebeest	hyena	lion
producer	primary consumer	secondary consumer	tertiary consumer
	herbivore	carnivore	top carnivore
trophic level 1 (T1)	trophic level 2 (T2)	trophic level 3 (T3)	trophic level 4 (T4)

- **Producers** are organisms such as plants, algae, and some bacteria, that produce their own food. They always start a food chain.
- **Consumers** are organisms that eat other organisms.
- Herbivores are animals that eat plants. They are also called primary consumers. They are the second link in the food chain.
- Carnivores are animals that eat other animals. They are also called secondary consumers. They are the third link in the chain.
- Top carnivores are animals that eat other carnivores.
- Omnivores are animals that eat both animals and plants, and are therefore both primary and secondary consumers.

Detrital food chains

These food chains start with dead material or waste. They involve two other feeding types:
- Detritivores are organisms that eat dead material such as dead leaves.
- Decomposers decay the bodies of other organisms, but they do not eat them.

Food webs

Food chains show animals eating only one food source. In reality almost all animals eat a wide variety of food. This means that animals in one environment appear in many food chains. One animal may link several food chains, forming a **food web**.

▲ Crown of thorns starfish feeding on coral

How scientists work

A case study: invading species

The Great Barrier Reef is a world-famous coral reef off the north-eastern coast of Australia. Divers and scientists noticed that the coral was decreasing in the 1980s. Scientists were asked to investigate the problem.

Step 1: hypothesis

First, the scientists made some simple observations and considered the problem. They then came up with an idea called a hypothesis, which might explain the cause of the problem. Scientists suggested that the coral was being overeaten by a carnivore, the crown of thorns starfish.

Step 2: testing the hypothesis

To test their hypothesis they needed to identify the food chains in this environment, and to collect some data about the numbers of the different organisms. This is one of the food chains that they worked out:

plankton → coral → crown of thorns starfish → wrasse

This shows that there is a delicate balance between the coral, the starfish, and its predator – the top carnivore, which was a fish called the wrasse.

▲ The wrasse, a predator of the crown of thorns starfish

Step 3: evidence supports the hypothesis

From their data on the numbers of these organisms, the scientists found that the wrasse was being overfished by humans. The result was that there was little predation of the crown of thorns starfish, and so it was overeating the coral. This was the cause of the decrease in the coral numbers.

Key words

food chain, trophic level, producer, consumer, food web

Questions

1 Write a food chain containing four organisms from the food web on the previous page.

2 A scientist has come up with a hypothesis. Explain why the next step is to collect data.

3 Other scientists often repeat the original scientist's experiments. Explain why this is done.

4 When researching the crown of thorns starfish feeding on the coral, scientists noticed that it had other predators. Use the Internet to:

   (a) Find out what those predators were.

   (b) Write a food web to show all these organisms.

   (c) Suggest what impact these other predators might have on the damage to the coral reefs.

Pyramids of numbers

Biologists studying food chains noticed that further along a food chain, there are usually fewer organisms. So there are fewer carnivores than herbivores.

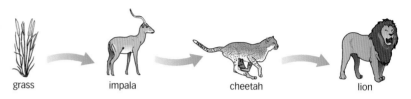

▲ A food chain from the African savannah

There were large numbers of small grass plants at the start of the chain, but only a few lions at the end of the food chain. Biologists used these numbers to plot a **pyramid of numbers**. This shows the number of organisms at each link in the food chain.

- The organisms are plotted as separate blocks.
- The size of each block is related to the number of organisms.
- The producer always goes at the base, with each successive link above it.

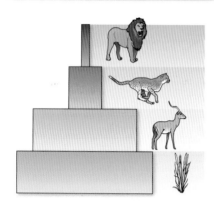

▲ A pyramid of numbers for the food chain on the African savannah

> **A** What does a pyramid of numbers show?

Unusually shaped pyramids of numbers

Sometimes the pyramid has a strange shape. For example, maybe one very large producer feeds many smaller consumers.

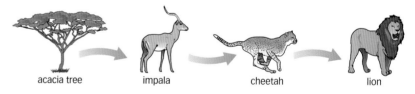

▲ A second food chain from the African savannah

When this is plotted as a pyramid of numbers, it loses its pyramid shape.

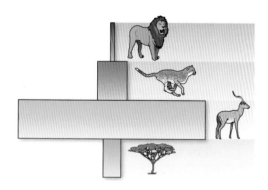

▲ A pyramid of numbers for the acacia tree food chain

Key words

pyramid of numbers, biomass, pyramid of biomass

> **B** Sketch a likely pyramid of numbers for the following food chains:
> (a) grass → caterpillar → blue tit → hawk
> (b) grass → impala → cheetah → fleas
>
> **C** Sketch a likely pyramid of biomass for the food chains in Question B.

Pyramids of biomass

These unusually shaped pyramids can be corrected by taking into account the mass of the organisms. To do this, biologists multiply the number of organisms at each link in the food chain by the dry mass of one organism. This gives them the **biomass** for that link in the chain. This can now be plotted as a **pyramid of biomass**.

How scientists work

A case study: lions under threat

Pyramids give us a picture of the state of the environment. If the pyramids change shape over time, this might suggest a problem in the environment. Scientists have been concerned about falling numbers of lions on the African savannah.

Step 1: analyse the data

Scientists have collected data over the last 60 years for lion food chains and plotted pyramids.

All the organisms have fallen in number, but particularly the lions. There were 500 000 lions in 1950, but only 20 000 in 2010.

Step 2: interpreting the data

The scientists identified a relationship between the falling numbers and human activity on the savannah. They suggested that the fall in numbers might be due to

- habitat destruction by humans to build towns, which reduces the numbers of all species
- hunting, which has a specific effect on the lion population.

Step 3: use the information to inform decisions

Because of these findings, people around the world who care about wildlife need to develop plans to protect the lion before it becomes endangered.

▲ A pyramid of biomass for the acacia tree food chain

Difficulties with food pyramids

In reality, constructing pyramids is not easy. Many organisms eat lots of things, so could be at different levels. Also, calculating dry mass involves killing and drying organisms.

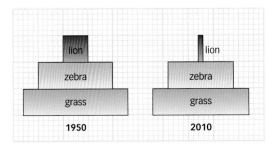

▲ Pyramids of numbers for an African savannah food chain, 1950 and 2010

Questions

1 How do scientists use food pyramids?

2 Construct a pyramid of biomass for the following food chain. Use graph paper, one small square to represent 100 kg.
   1 000 000 grass plants (0.1 kg each) → 100 zebra (300 kg each) → 4 lions (250 kg each)

3 Explain why scientists plotted pyramids of numbers rather than pyramids of biomass for the lion food chains.

Key words

energy, egestion, excretion, efficiency

Energy

Food chains and food webs show not only the flow of materials in biomass from one organism to another, but also the flow of **energy**. All living things need energy to stay alive.

Energy enters food chains and webs when plants capture sunlight energy. This means that all other organisms rely on plants to start the chain. Energy flows through the food chain, and leaves it as heat or in waste materials. There must be a continual supply of energy into the food chain to maintain life. However, some of the energy leaves the food chain at each link. The energy in waste or in uneaten parts of plants and animals can also be used to start new food chains.

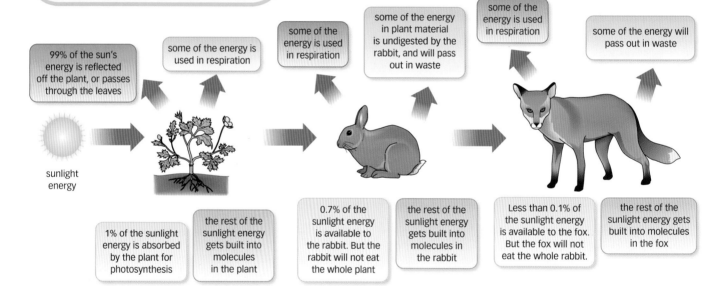

▲ Energy flows through a food chain. Some energy is transferred out at each stage.

A In the food chain above, how much of the Sun's energy reaches the fox?

How is energy transferred out of the food chain?

- Not all the food eaten by animals is digested. Some passes straight through the body undigested and comes out in droppings. This transfers energy out of the food chain by **egestion**.
- Every organism in the food chain uses some energy for respiration. This process releases energy for the animal's movement and heat to warm the body.
- Some energy is also lost by **excretion** in excretory products such as sweat and urine.

Because energy leaves at each link in the food chain, there is very little energy left for the organisms at the end of the chain. This keeps food chains short. Most land food chains do not have more than five organisms. This is also why food pyramids tend to have fewer organisms and less biomass at the top, because there is less energy than at the base.

B State three ways in which energy is transferred out of the food chain.

C Name the source of energy for (a) a plant (b) a herbivore.

Energy efficiency in farming

Modern farming needs to produce food as cost-effectively as possible. Farmers need to minimise energy losses from the food chain to get the best yield. Science tells us the following, which helps the farmer:

- The shorter the food chain, the less energy is transferred out. Farm animals tend to be herbivores. This is more energy **efficient**. If farmers raised carnivores for us to eat, they would need to farm large numbers of herbivores to feed them.

- It is possible to calculate the efficiency of energy transfer. Farmers might do this to compare methods of rearing animals. To do this, they use the formula:

$$\text{efficiency} = \frac{\text{energy built into organism}}{\text{energy available}} \times 100\%$$

- Less energy is transferred out from animals if they use less heat to keep themselves warm. In intensive farming, larger animals may be kept indoors in a barn for this reason, or smaller animals like chickens are often reared in cages. However, it is important to consider animal welfare issues; some battery conditions are inhumane.

- Energy can also be kept in the food chain by reducing animals' movement. Again, this is achieved by keeping animals indoors or in cages.

- Farmers and growers need to minimise the energy that is transferred out to pests if they eat crops, or to weeds that compete with their plants for light.

▲ Battery farming reduces energy losses

Questions

1 Name the process used by plants to convert light energy to chemical energy.

2 What happens to most of the Sun's energy when it hits a plant's leaf?

3 Explain, in terms of energy, why there are far more zebras than lions on the African savannah.

4 On free range farms the animals are allowed to roam freely outdoors. Explain why this results in produce which is more expensive.

E

↓
A*

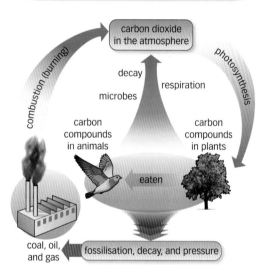

▲ The major steps in the carbon cycle

A What is the impact of burning fossil fuels on the carbon cycle?

B How might planting trees balance the carbon given out by your activities in daily life?

Nutrient cycles

Elements pass between the living world and the non-living world – air, water, soil, and rocks – in a constant cycle. As living things grow they take in elements, including carbon and nitrogen, to use in their bodies.

Carbon in the living and non-living world

Carbon is one of the most common elements in living things. All of our major molecules, including carbohydrates, proteins, fats, and DNA, contain carbon. The process of carbon moving between the living and non-living world is called the **carbon cycle**.

• Carbon is present in the atmosphere as carbon dioxide.

• Carbon dioxide is absorbed by plants, and built into carbohydrate molecules such as sugars. This happens during **photosynthesis**.

• The plant uses some of the sugars to make other molecules such as cellulose, fats, and proteins, which it uses to grow.

• Plants are eaten by animals and so these carbon compounds can pass into animals and become part of their bodies.

• Both plants and animals **respire**. This returns some carbon dioxide back to the atmosphere.

• When plants and animals die, soil bacteria and fungi known as **decomposers** digest their bodies in the process of **decay**, and carbon dioxide is released back into the atmosphere through respiration.

• Not all plant and animal bodies will decay. Some are buried under layers of silt and over millions of years begin to fossilise.

• This forms fossil fuels such as coal, oil, and gas.

• Humans extract fossil fuels and burn them to release energy.

• Burning fossil fuels releases the carbon that was stored millions of years ago as carbon dioxide.

• In a stable environment the amount of carbon dioxide released should approximately equal the amount absorbed.

Useful microbes in the cycle

Microbes are involved in the decay of dead plants, animals, and animal waste. Two main groups of microbes are involved – bacteria and fungi. The process of decay releases nutrients such as carbon back into the environment. These nutrients can then be reused by plants to grow.

This process of decay keeps the carbon cycle turning. Decay happens more at certain times of the year, especially during autumn. This is when the microbes have the right conditions to survive and respire:

- food (dead plants or fallen leaves)
- oxygen
- a suitable temperature
- moisture
- suitable pH.

Where conditions are poor for the microbes, the rate of decay is slow. For example, in waterlogged soils there is a lack of oxygen for the microbes. In acidic soils the pH is too low for the microbes.

Locked-up carbon

Carbon built into the bodies of plants or animals may stay there for millions of years.

In the oceans carbon dioxide dissolves. There are many invertebrate animals with shells to protect their bodies. These shells are made of compounds such as calcium carbonate. The animals have incorporated carbon gained from digesting plants into their shells. When they die, their shells are difficult to decay, and they sink to the bottom of the ocean. The fragments are compressed together to form rocks such as limestone and chalk. Because they store carbon in this way, the oceans are called **carbon sinks**.

Eventually, the carbon may be released back into the atmosphere. Some of these limestone rocks become exposed above the sea. Gradually, the limestone and chalk are eroded by weathering, and the carbon is released as carbon dioxide into the air. Volcanoes also release carbon dioxide into the atmosphere.

C Use your knowledge of microbes to suggest why the carbon cycle slows down during the winter.

D Why is autumn a good season for the process of decay?

▲ The chalk cliffs at Dover contain locked-up carbon

Questions

1 (a) Name the process by which carbon enters living organisms.
  (b) Name two processes by which carbon is released from living organisms.

  E

2 Describe the key role of microbes in the carbon cycle.
  C

3 Describe how the actions of humans are leading to an imbalance in the carbon cycle.
  A*

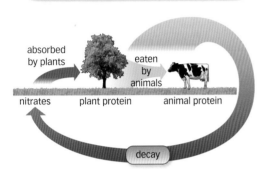

Nitrogen in living things

As well as carbon, another element that is cycled between the living and non-living world is nitrogen. Nitrogen makes up 78% of the air. It is important to living things because it is used in two major groups of molecules – proteins and DNA. Nitrogen is a very unreactive gas, and most organisms cannot make use of it as an element. They can use nitrogen in compounds such as nitrates.

Nitrogen in the living and non-living world

Nitrogen moves between the living and non-living world in the **nitrogen cycle**.

- Nitrogen exists as **nitrates** in the soil, dissolved in water.
- Dissolved nitrates are absorbed by plant roots.
- The nitrates are used to build compounds such as proteins in the plant.
- These nitrogen-containing compounds pass through food chains to animals.
- Nitrogen compounds (such as proteins) in dead plants and animals are broken down into nitrates for absorption by plants.

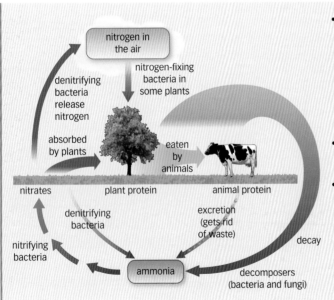

- Decomposers return the nitrates in plants and animals to the soil in two ways:
  - animal wastes, urea and faeces, contain nitrogen, and are converted to ammonia
  - when plants and animals die and decay, their proteins break down, releasing ammonia.
- Nitrifying bacteria in the soil convert the ammonia to nitrates to complete the cycle.
- There are also denitrifying bacteria in the soil that break down nitrates, releasing nitrogen into the air.

- There are three ways that nitrogen gas can get back into the soil:
  - Lightning releases enough energy to cause nitrogen in the air to react with oxygen to form nitrates, which wash into the soil.
  - There are nitrogen-fixing bacteria in the roots of plants such as peas, beans, and clover. These plants are called **legumes**. The bacteria combine nitrogen in the air with oxygen to make nitrates, which the plants can use.
  - Chemists are able to make ammonia industrially, using the Haber process.

Microbes: friend or foe?

In the nitrogen cycle there are some bacteria and fungi that help form nitrates, which plants can use, from unreactive nitrogen.
- Nitrifying bacteria help dead material decompose into ammonia, and convert ammonia into nitrates.

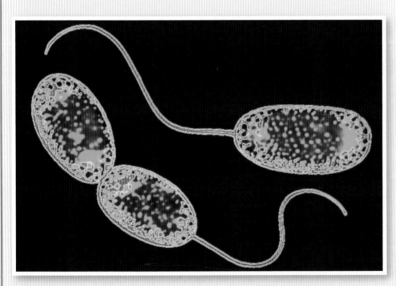

▲ Electron micrograph of bacteria that help convert ammonia into nitrates (× 11 000)

Other bacteria do the opposite, and tend to reverse the cycle.
- Dentrifying bacteria break nitrates down, and release nitrogen into the air.

Key words

nitrogen cycle, nitrates, legume

A Explain why decay is important in the nitrogen cycle.

B If farmers remove crops from the soil, there are no leaves to decay into the soil. How can farmers replace the nitrates taken from the soil?

Questions

1 State the percentage abundance of nitrogen in the air.

2 What molecules in plants and animals are nitrates used to build?

3 Describe the role of decomposers in the nitrogen cycle.

4 An organic farmer does not use chemical fertilisers on their farm. How could they fertilise their crops?

5 Explain why medieval farmers grew clover on each field one year out of every four.

↓ E

↓ C

↓ A*

Learning objectives

After studying this topic, you should be able to:

✔ know that organisms compete with each other for resources, and that this can affect their distribution

Key words

resource, competition, population, niche

Did you know...?

Humans are probably the greatest competitors of all time. This has led to many other species losing out. Dodos and Tasmanian tigers have both become extinct because of humans.

Exam tip OCR

✔ When two populations compete with each other, the number of individuals in each population is reduced.

A Why do farmers and gardeners remove weeds from around their plants?

B Explain why bluebells only grow in the spring in a British woodland.

What do plants and animals compete for?

There is a limited supply of **resources** for plants and animals. Nearly all living things are locked in a battle to get enough materials and other resources to survive. This fight for resources is called **competition**. The availability of resources can affect the distribution of an organism – where it lives.

Plants compete for
- light
- space
- water
- minerals
- carbon dioxide.

Animals compete for
- food
- space or territory
- mates
- water
- shelter.

A **population** is the number of organisms of a particular species in a named area. For example, the woodlice living under a stone are a population. If conditions are good and the woodlice reproduce, the population gets larger. If conditions are bad and some woodlice die out, the population gets smaller. Competition affects the size of populations.

Competition between plants

Carbon dioxide: plants need this for photosynthesis. The air contains only 0.03% carbon dioxide. The massive canopy of tree leaves absorbs carbon dioxide, so there is less available under the tree for ground plants.

Light: the energy supply for photosynthesis. The tree leaves absorb some light, and not much light passes through, so it is too shady under a tree for most plants to grow.

Space: the tree roots take up most of the space in the ground, leaving little room for ground plants.

Water: used in photosynthesis and to cool the plant. The large tree will absorb most of the water in the soil, leaving little for the smaller ground plants.

Soil minerals: needed to keep the plant healthy. These are absorbed in water through the roots. Again, the tree can absorb far more minerals than the small plants.

▲ Plants do not move around, so they can only live in places where resources are available. If they cannot compete with other plants for these resources, they will not be able to survive there.

Competition between animals

Animals usually compete with each other for food. For example, some coastal birds compete for the same food supply and they must find a solution for the species to survive together.

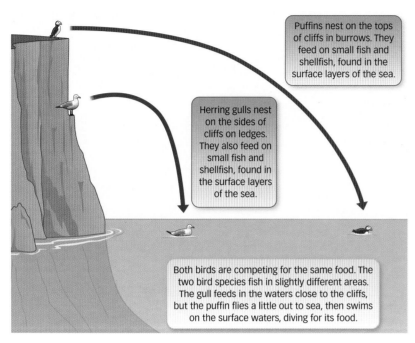

Puffins nest on the tops of cliffs in burrows. They feed on small fish and shellfish, found in the surface layers of the sea.

Herring gulls nest on the sides of cliffs on ledges. They also feed on small fish and shellfish, found in the surface layers of the sea.

Both birds are competing for the same food. The two bird species fish in slightly different areas. The gull feeds in the waters close to the cliffs, but the puffin flies a little out to sea, then swims on the surface waters, diving for its food.

▲ Competition between the puffin and the seagull

The niche

The two species of bird in the diagram above, the puffin and the herring gull, have developed different lifestyles in order to survive. The unique way an organism has of surviving in its environment is called its **niche**. The two birds here have similar niches, in that they live on cliffs and feed on the same food. But they feed in slightly different areas of the sea, which means that they have slightly different niches.

Types of competition

Interspecific – this is between organisms of different species. Red and grey squirrels compete for the same food, for example.

Intraspecific – this is between groups of organisms of the same species, such as two troops of baboons on the African plains. This type of competition is often more significant because the needs of the two groups are identical.

C Name the most common resource animals compete for.

D If humans fished the coastal surface waters in the diagram, explain what would happen to the population of puffins.

Questions

1 Explain why plants need light to survive, but animals do not.

2 Explain what will happen to the ground plants in a woodland when a large tree dies and falls.

E

3 Red squirrels are a native species in the UK. In around 1900 a close relative, the grey squirrel, was introduced. The grey squirrel is a better competitor. Describe the effect the grey squirrel's introduction had on the red squirrel population.

C

4 Describe how the niche of the giraffe and zebra are different, which allows them both to survive together on the African savannah.

A*

Learning objectives

After studying this topic, you should be able to:

✔ know that organisms live together, and affect each other's survival

✔ know the features of predators and prey, and how those features aid their survival

Key words

relationship, predator, prey, mutualism, parasitism

A Think of five predators. Name at least two features of each that make them effective.

B Explain how the camouflage of a zebra helps it avoid predators in its environment.

Predator and prey relationships

Species rarely live alone. There are a number of ways in which organisms interact with one another. These interactions are called **relationships**.

Predator

Prey

Characteristic features:	Characteristic features
• binocular vision to judge size and distance when hunting • strength and speed to chase prey when hunting • camouflage to prevent detection during stalking • good senses to detect prey, such as hearing and smell • large teeth and claws to hold and kill prey • hunt either in a pack, or alone using stealth • have small numbers of young (insufficient prey to support more).	• eyes on the side of the head to give good all-round vision • speed to escape predators • herd lifestyle to reduce chance of individuals being killed or singled out by predators • some have stings or poison. • camouflage (cryptic colouration to break up outlines, warning colouration using colours of other dangerous organisms, or mimicry, copying the body patterns of predators) • have large numbers of young (as many will be killed) • young born simultaneously (so only some are killed).

Predator and prey populations

In a predator–prey relationship, the size of each population directly affects the size of the other.

A study using data from fur trappers in Canada on the predator–prey relationship between the lynx and the snowshoe hare was carried out. The lynx was the main predator animal in northern Canada. So biologists could see clear links between the populations of the two animals. The population numbers were plotted as a graph.

When the snowshoe hare population increased, there was far more food for the lynx. This led to an increase in the lynx population slightly later. This bigger lynx population then needed more food, and so more hares were eaten. The hare population fell. The cycle repeats.

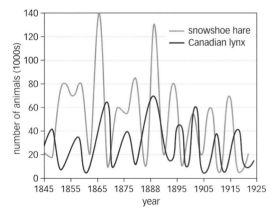

▲ This famous predator–prey study was carried out in Canada between 1845 and 1937

▲ The predator population changes are slightly out of phase with the prey, as it takes time for the predator numbers to change in response to the prey number

Other types of relationship

There are other relationships between organisms where the populations affect each other differently. Two common relationships are:

- **mutualism** – here both organisms gain some benefit
- **parasitism** – here one organism benefits at the cost of the other.

The oxpecker (an example of a cleaner species) is a small bird which eats the ticks, fleas, and other parasites from the skin of buffalo. This is an example of mutualism. Another would be pollination of plants by insects.

A parasite feeds on or in a second animal, called the host, which gains nothing from the relationship, or may be harmed by it. Fleas are parasites that live on the skin of their host and feed on its blood. Tapeworms are parasites that live in the gut of the host, and feed on the food inside the host's gut.

Plant relationships

Bacteria are involved in many mutualistic relationships, including with plants. Nitrogen-fixing bacteria live in the root nodules of leguminous plants, the pea and bean family. The bacteria take nitrogen from the air and convert it to nitrates in the roots – they fix the nitrogen. The plant gains the nitrates, and the bacteria gain food from the plant cells.

▲ Lynx with snowshoe hare

> C When snowshoe hare numbers fell, describe what happened to the numbers of lynx.
>
> D Explain why it was important in this study that there were no other predator animals in the environment.

▲ Cape buffalo with a yellow billed oxpecker

▲ A cat flea between cat hairs (×70)

Questions

1 When rabbits were accidentally released in Australia, they had no natural predators. Suggest why this became a problem for farmers.

2 Use your knowledge of how scientists work to suggest why the lynx and hare study took place over several years.

3 Many flowers are pollinated by butterflies. The butterflies feed on the nectar from the flowers. What kind of relationship is this?

4 Suggest some of the advantages to prey populations of having a natural predator.

Learning objectives

After studying this topic, you should be able to:

- ✔ know that adaptations of organisms help them survive
- ✔ identify and explain adaptations of animals

What are different environments like?

Conditions in different environments can vary greatly. Deserts are hot and dry environments; the Arctic is cold. There are animals that can survive in each of these environments, but they must be adapted to survive. An **adaptation** is a feature of an animal's body which helps it to live in its environment, making it better able to compete for limited resources.

Adaptations to cold environments

Adaptation	How this aids survival
small ears	reduces the surface area of the ear, and so less heat is transferred to the environment
thick white fur	insulates the body against the cold, and camouflages the bear
sharp teeth	to kill prey
strong legs long legs	contain large muscles that contract so the bear can run on land or swim in water
big feet with fur on the soles	spread the load of the animal on the snow or ice; fur helps grip and helps insulate against the snow

Adaptation	How this aids survival
claws	for killing and holding prey
blubber below the skin	a thick layer of fat that insulates against heat transfer to the environment; the stored fat can also be used for respiration to generate heat
large body size	reduces the relative surface area, and so reduces heat loss
fins	balance the whale during swimming
muscular tail	contains large muscles that contract to generate movement during swimming to catch prey

Adaptations to hot and dry environments

Adaptation	How this aids survival
dry dung and concentrated urine	reduce water loss in waste
hump of fat	fat is stored in one place, which reduces all-round insulation; fat can be broken down to release water (valuable where water is scarce)
nostrils which close	prevent breathing in of sand
bushy eyelashes	stop sand entering the eyes
body tolerance to temperature changes	does not need to sweat so much when hot
long legs	lift body off hot sand

Adaptation	How this aids survival
large feet	spread the load, stop the camel sinking into the sand
thin fur	less warm air is trapped, reducing insulation
large ears	lose heat by radiation, and used to fan the body
large body size	can knock over plants and shrubs for food
wrinkled skin	increases the surface area from which to lose heat
trunk	allows the elephant to suck up water to drink, and to spray water over the body to cool itself
large feet	spread the load, stop the elephant sinking into the mud

Coping with extreme temperatures

Organisms called extremophiles can survive in extremes of temperature, demonstrating adaptations that enable them to survive.

Biochemical adaptations

1. Bacteria living in hot water springs and geysers in volcanic areas can survive temperatures above 80 °C. Their enzymes work well at these high temperatures. Normal enzymes would be denatured.
2. Some organisms have to withstand extreme cold. Periwinkles living on the beach produce glycerol as an antifreeze in winter. Carrots in frosty soils produce a protein-based antifreeze.

Physiological adaptations

Penguins living on ice have a special blood flow system called a counter-current exchange system.

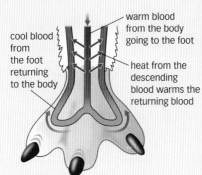

warm blood from the body going to the foot

cool blood from the foot returning to the body

heat from the descending blood warms the returning blood

Questions

1. A close relative of the elephant is the woolly mammoth. These lived in cold environments. Suggest two features that differ from those of the elephant and helped the mammoth survive.
2. Explain how the features you identified in Question 1 would aid the mammoth's survival.
3. The distribution of camels is limited to the desert. Explain why the camel is well adapted to life here.
4. Explain why slow-moving camels are not well adapted to live in the community of animals on the African savannah.

↓ E

↓ C

↓ A*

B Sort the adaptations of the elephant and camel into two lists – those that help them survive hot conditions, and those that are not related to the heat.

Adaptations affect distribution

A well adapted animal survives well in its environment. It is fit for its job. Elephants are well suited to the grassy African savannah, so there are large numbers of elephants there. They cannot necessarily survive well elsewhere. They are **specialists**. Elephants are not adapted to survive further north in Africa, where there are deserts, because they cannot survive long periods without water, and there would not be sufficient food for them. So the distribution of African elephants is limited to the savannahs. Other organisms are able to live in a range of habitats. They are **generalists**, but are easily out-competed.

Exam tip   OCR

✓ When looking for adaptations, notice particular features of an animal, and try to suggest how it uses the features to survive.

Learning objectives

After studying this topic, you should be able to:

✔ know that plants show adaptations to hot and cold environments

✔ understand that the adaptations of a plant determine where it can grow

Plants are adapted, too

Plants are found in all sorts of environments. In order to survive they need to be adapted just as well as animals, to be better able to compete for limited resources. Here are some plant environments:

- northern pine forests – cold for much of the year; all the water is often frozen for some months
- temperate forests – as in the UK, where the winters are cold and dark but the summers are sunny and quite moist
- deserts – where the ground is very dry and hot.

Adaptations to different environments

Northern pine forests	Temperate forests	Deserts
Pine tree	Oak tree	Cactus
Conditions	**Conditions**	**Conditions**
Water is scarce as for much of the winter the water freezes in the soil and cannot be absorbed by roots.	Cold and not much light in winter; sunny and moist in summer.	Very little water in the sandy soil; very low rainfall. Air is hot and dry.
Adaptations	**Adaptations**	**Adaptations**
Leaves are reduced to **needles** to reduce the **surface area**; this reduces water loss. Thick wax on the surface of the leaf also reduces water loss. Pine needles have fewer pores from which water can be lost.	Leaves fall from the trees in the autumn – there is not much light for photosynthesis, so leaves have no function in the winter. Leaves have a larger surface area to make the most of the summer sun. Wax on the leaf surface is thinner as there is plenty of water in the soil for most of the year.	Leaves are **spines**, which reduces their surface area; this reduces water loss. Rounded shape reduces surface area to volume ratio. Spines are less likely to be eaten by animals. Thick wax cuticle reduces water loss. Shallow root system to cover great areas, to absorb water when it does rain. Stems are swollen to store water, and some have grooves so they can expand. Green stem for photosynthesis.

Key words

needles, surface area, spines

A What is the advantage to the oak tree of its large green leaves?

B Name the advantage to the pine tree of reducing its leaves to needles.

C Suggest a disadvantage of reduced leaves to the pine tree.

D The pine tree and the cactus both show similar adaptations of their leaves. Explain why the same adaptation helps plants survive in two different environments.

Thermogram of a polar bear, showing minimal heat loss from the ears

Stressful environments

The environment is always changing. If these changes are large, it can mean that plants and animals have to cope with environmental stress. Such stresses include changes in temperature and water availability.

Dealing with stress

Surface area to volume ratio

Surface area to volume ratio varies in plants and animals. Animals in hot environments often have an increased surface area to volume ratio, with structures such as large ears in elephants promoting heat loss. A polar bear in a cold environment, however, has small ears and a reduced surface area to volume ratio in order to help minimise heat loss.

Plants such as pine trees and cacti in areas where water is scarce may have a reduced surface area to volume ratio, reducing the surface area over which water can be lost.

Animal behaviour

Animals use behaviour patterns to cope with environmental stress such as temperature changes and water loss:

- Migration – some animals move to more favourable conditions. Canada geese migrate south for a milder winter.
- Hibernation – some animals hibernate when the temperature falls, slowing down their life processes and entering a dormant state until conditions improve.
- Sun basking – some cold blooded animals, like crocodiles, lie in the morning sun to warm their bodies after the cold night. This allows them to become active. In hot weather they cool off in the water.
- Water storage – some animals, like camels, can drink and store huge quantities of water when they find it.

Exam tip **OCR**

✓ Think about how a plant is adapted to its environment, and how its adaptations restrict its distribution.

Questions

1 Explain why the dandelion isn't well adapted for living in the desert.

2 Suggest a reason why flowering plants tend to reproduce more in the warmer summer months.

3 The acacia tree has a very long tap root. Explain how this adaptation helps it to survive on the African savannah.

4 Explain why it is important that a crocodile keeps moving between the land and the water during the day.

↓ C

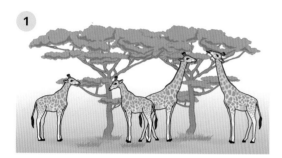

Evolution

The world is full of millions of different species. Where did they all come from? This has puzzled biologists for many years.

- Why are there so many different species?
- How do species change or adapt over time?
- How did all the different species form?
- Why are some species closely related to each other?

Most biologists believe that an idea known as evolution best answers all four questions.

There are many different habitats in the world. In each habitat there are organisms that are well adapted to survive. This results in lots of species. But the habitats of the world are constantly changing. The organisms must also change to survive.

Evolution is the gradual change of species over time. This idea suggests that gradually one type of organism, called the ancestor, might change over many generations into one or more different species. This generates lots of different species over time. But how do organisms change?

Natural selection

The biologist Charles Darwin suggested an idea to explain how one species can change into another. This is now widely accepted by most biologists and is called **natural selection**. This theory can be used to explain how the giraffe's long neck evolved. There are four major steps in the theory.

1. Large populations

- Most species produce lots of offspring. This should cause a massive population growth for every species. One original pair of giraffes would produce millions of giraffes over a few hundred years.
- But the population seems to remain roughly the same. Why?

2. Survival

- If all the organisms die, the species becomes **extinct**.
- Their survival is affected by changes in the environment. Some die from disease. Some starve. Some are eaten by predators. Some cannot find a mate.

3. The fittest

- In any population there is a lot of variation, caused by chance mutations in genes. Some of the variations will be successful adaptations and are an advantage.
- Some giraffes have longer necks than others.
- When food lower down is scarce, without the advantage of a longer neck a giraffe dies.
- Giraffes with longer necks reach leaves higher in the tree. This is known as 'survival of the fittest'.

4. Passing on the advantage

- The surviving giraffes are the only ones that reproduce.
- Their offspring inherit the advantage. The gene for long necks has been passed on.
- Over many generations the number of giraffes with long necks increases.
- The result is that the giraffe species has changed to become one with long necks.

A Define evolution.

B Darwin used the term 'survival of the fittest'. Explain what this means.

Forming new species

Sometimes two different groups of the same species gradually change over time in different ways. This may form new species. Biologists have explained the process in three steps.

1. The populations become isolated by some kind of barrier. The barrier may be geographic (such as a mountain range, river, or the sea) or behavioural (such as reproductive seasons not coinciding).
2. Over time, each isolated group of the population evolves differently in different conditions on each side of the barrier. The longer they are isolated, the more different they become.
3. The two sub-populations have changed so much that they can no longer interbreed. They have formed separate but closely related species.

▲ The African elephant and the Asian elephant are closely related species, which both evolved from a common ancestor many years ago when they were separated by great distances.

Key words

evolution, natural selection, extinction

Questions

1 State what often happens to organisms that do not evolve when their environment changes. ↓ E

2 Pet shops sell white and brown rabbits. White rabbits are easily seen by foxes. Use Darwin's theory of evolution to explain why white rabbits are rare in the wild. ↓ C

3 Lions and tigers evolved from a common ancestor. Explain how moving to different habitats resulted in the formation of these two different species.

4 When explorers discovered Australia they were amazed by how different the animal species were. Explain why Australian animals are so different from European ones. ↓ A*

▲ Charles Darwin aged 40

▲ Alfred Russel Wallace

A It was important to Darwin that Wallace had made similar findings. Why?

▲ Jean-Baptiste Lamarck

How did Darwin gather his evidence?

Charles Darwin travelled the world on the ship HMS *Beagle*. He made notes on the different types of plants and animals he saw. This evidence helped him develop his ideas on evolution – how the different species might have arisen. His ideas explained all the observations he made while travelling.

Darwin published his ideas in 1859 in a book called *On the Origin of Species by Means of Natural Selection*. Many people were horrified. Other theories were available, and people did not like the idea that humans could have evolved from apes.

Most people, including many scientists, based their ideas about where plants and animals came from on the Bible. Darwin's theory of evolution seemed to contradict the Bible, and therefore God. Over time this theory has been discussed and tested by scientists and is now widely accepted.

> As new discoveries have been made, such as the knowledge of genes, the theory of natural selection has devleoped.

Alfred Russel Wallace

Alfred Wallace was a Welsh biologist who worked at the same time as Darwin. Although he collected a lot of evidence, he was not as well-known as Darwin. He came up with the same idea as Darwin to explain how species were formed.

An alternative idea: Lamarck's theory

Jean-Baptiste Lamarck was a French biologist who worked before Darwin. He would have explained the evolution of the giraffe as follows:

- In a giraffe population, some giraffes want to feed off the leaves high on a tree.
- The giraffes stretch their necks, becoming more successful.
- They pass their longer neck on to their offspring.

The ideas put forward by Lamarck are not accepted today. Evidence shows that the only changes in organisms that can be passed on to offspring are changes in genes, not changes acquired during the organism's lifetime. Lamarck's theory cannot be accepted because it does not take into account the evidence of inheritance by genes.

Living fossils

There are some organisms that have not changed over millions of years. This is possibly because their environment has not changed. These species are called living fossils. Examples include the fish called coelacanths, sharks, and crocodiles.

Examples of evolution

1. The peppered moth (*Biston betularia*)

Perhaps the best known example of evolution in action involves a moth called the peppered moth.

The pale moth was well camouflaged against the light bark of trees. During the 1800s trees in industrial areas became covered in soot particles, and the bark became much darker. The moths were no longer camouflaged, and they were eaten by birds.

A mutation occurred in some moths, making them much darker in colour. The darker moths now had the advantage of camouflage.

Over the next 50 years, the dark variety became more common.

Today, in cleaner areas the light form of the moth is more common again. In industrial areas the dark form is still the more common form.

> B Use Darwin's theory of natural selection to explain the steps involved as the moth population changed to the darker form.

2. Antibiotic resistance in bacteria

Antibiotics are drugs that kill bacteria. If a bacterium has a mutation which makes it **resistant** to the antibiotic, it will survive and reproduce. The new bacteria will all be resistant to the antibiotic.

3. Warfarin resistance in rats

Warfarin is a poison that has been used for many years to kill rats. A mutation occurred in some rats, which made them resistant to the warfarin. They then reproduced, passing on the resistance genes. Over time, the population of rats in the UK is becoming resistant to warfarin. The rat population is evolving.

▲ Light and dark varieties of peppered moth

Key words

resistance

Exam tip OCR

✔ When comparing the ideas of two different scientists, list the key points of each one and say what is similar and what is different.

Questions

1 Explain why biologists look for more than one piece of evidence for their theories. ↓E

2 Scientists are continually developing new antibiotics. Explain why new antibiotics will only be effective for a few years. ↓C

3 The ideas of both Darwin and Lamarck can be used to explain the evolution of the giraffe's long neck. Explain how each of these scientists would have accounted for the giraffe's long neck. ↓A*

Learning objectives

After studying this topic, you should be able to:

✔ know that the human population is increasing

✔ understand that this increase is unsustainable and leads to pollution

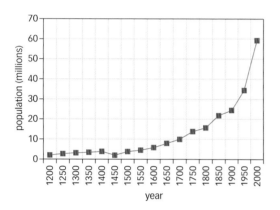

▲ The human population of the UK has been rising fast

Key words

exponential growth, pollution, sustainable, pollutant

Did you know...?

Finland is a Green Champion. It was the first country to introduce a carbon tax. The government imposes a tax on companies which release carbon dioxide into the air.

The human population

The number of humans on the planet has been increasing. because the birth rate exceeds the death rate. But the increase is not steady. Biologists have studied the growth of the human population and found some interesting points.

1000 years ago, the population was stable:
- not many people to reproduce
- limited food supply.

Between 1600 and 1900, there was a steady increase:
- better farming methods
- improving hygiene.

After 1900, there was a dramatic rise in the population:
- improved diet
- improved healthcare
- improved hygiene
- lower infant mortality (death rate).

This type of population growth is called **exponential growth**. This means that the more individuals there are in a population, the faster the population growth.

> **A** Families had large numbers of children before 1900. Suggest why.
>
> **B** What must happen to the birth rate and death rate to keep the population fairly constant?

The population explosion

The massive increase in human population seen in the UK is repeated in most countries in the world. This increase in the world's population has a number of effects:
- There is a shortage of food in some countries.
- More land is being used for building and farming.
- More **pollution** is being produced.
- The world's resources (such as fossil fuels and minerals) are being used up too fast.

This growth is not **sustainable**. The impact of so many humans on the environment is harmful.
The global effect of population growth is exaggerated by figures from developed countries. Despite having a smaller total population, they use more resources per person, creating more pollution.

Human influences on the environment

These can be classed into two major areas:

Agriculture

- use of fertiliser
- use of pesticides
- loss of habitat
- deforestation
- monoculture
- animal waste.

Towns and industries

- loss of habitat
- quarrying and extraction of raw materials
- dumping of wastes
- production of toxic chemicals
- sewage.

The development of towns and industries removes large areas of Britain's natural woodlands. Agriculture and industry also produce a number of **pollutants**.

Air pollution

Pollutant	Source	Effect on environment
sulfur dioxide and nitrogen oxides	burning (combustion) of fossil fuels	Dissolves in rain to form acid rain: • damages plant leaves • acidifies lakes • changes minerals available in water supplies • causes bronchitis.
carbon dioxide	burning (combustion) of fossil fuels	Dissolves in rain to form acid rain (see above). A greenhouse gas which keeps more heat in the atmosphere, leading to global warming.
CFCs	aerosols and refrigerator coolants	Destroy the ozone layer in the upper atmosphere, allowing more ultraviolet radiation through and contributing to skin cancers.

Water pollution

Pollutant	Effect on environment
untreated sewage	Bacteria can cause disease, e.g. typhoid and cholera. Nitrates in the water kill fish.
detergents	Kill organisms in the water and affect food chains.

▲ Biologist collecting water samples for testing, to directly measure pollution levels. This gives very accurate measurements of pollution, but only for the time at which the sample is taken.

Questions

1 State why the human population explosion is not sustainable.

2 Biologists say that the use of resources must be sustainable. Describe what this might mean.

 E

3 Recycling helps improve sustainability. Explain why this is the case.

4 Use the information on these pages to explain why the increase in the human population is leading to more pollution of the air.

 C

5 Everybody wants the latest mobile phone. However, there is a cost to the environment in buying such products. Identify the effects on the environment that producing mobile phones may have.

 A*

A What are the principal pollutants that cause acid rain?

B How is acid rain formed?

C What are the sources of the acidic gases?

Key words

indicator species

Did you know...?

Your carbon footprint is the amount of greenhouse gas given off in a period of time. Your carbon footprint is about 9.675 metric tonnes of carbon dioxide released per year.

D How are indicator species helpful to biologists?

Two serious consequences of pollution of our environment are acid rain and global warming.

Acid rain

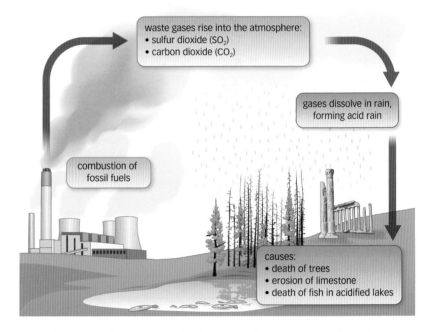

waste gases rise into the atmosphere:
• sulfur dioxide (SO_2)
• carbon dioxide (CO_2)

gases dissolve in rain, forming acid rain

combustion of fossil fuels

causes:
• death of trees
• erosion of limestone
• death of fish in acidified lakes

▲ Acid rain can fall many hundreds of miles from the place where the pollution was produced

Global warming

A number of pollutants contribute to a process called global warming.

• Burning fossil fuels releases greenhouse gases such as carbon dioxide.
• Greenhouse gases allow the Sun's radiation in, but prevent heat from being radiated away from the Earth.
• The atmosphere acts like a blanket, keeping the Earth warmer.

Pollution and biodiversity

When pollutants enter the environment, they will have an impact on the number and types of organisms that can survive. Generally, in more polluted areas fewer species can survive. In some cases particular species will be the main or only survivors, and may even be adapted to cope quite well with the pollution. The presence, or absence, of these species shows biologists that the area is polluted. They are called **indicator species**.

Indicator species in water

Rat-tailed maggots are adapted to survive in water with very little oxygen. They have a long, tail-like tube which is hollow. It acts like a snorkel, allowing the maggot to breathe in air containing oxygen from above the polluted water. Other species indicating polluted water are sludge worms and the water louse. Mayfly larvae are found in clean water.

Lichens: another indicator species

Burning fossil fuels releases many chemicals into the air, including sulfur dioxide. This causes air pollution, which reduces the variety of organisms that can survive in the area. Lichens are one group of living things that act as indicator species of air pollution.

◀ Pollution reduces the variety of organisms that can live in an area

Some lichens can cope with high levels of pollution, and are found in cities. Other lichens cannot grow there, and are only found in areas with clean air away from cities and motorways.

> Lichens are great indicators of the level of air pollution in an area, as they live in the environment for a long time. However, lichens are not present in all the areas we want to test.

Questions

1 Name an organism you might find in polluted water.

2 Describe an experiment you could carry out to show how lichens can be used to indicate the levels of pollution as you move out of a city.

↓ C

3 The carbon footprint is the amount of greenhouse gases given off in a period of time. Explain how Britain might reduce its carbon footprint.

↓ A*

4 What advantages are there in using living organisms to detect pollution?

▲ Rat-tailed maggots are adapted to survive in polluted water

▲ These lichens are indicators of (a) polluted air (b) moderate pollution (c) clean unpolluted air

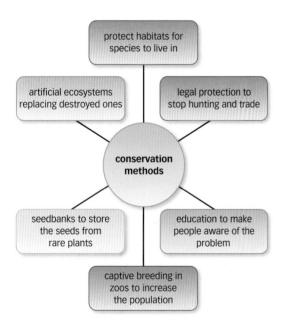

▲ Conserving endangered species

What is sustainable development?

Sustainable development means using resources for human needs without harming the environment. If these resources are taken from our environment without being replaced, this use is unsustainable.

Wildlife in danger

Extinction

As the human population has grown, more resources have been taken from our environment, damaging it. Some species could not adapt to these changes, did not survive, and became extinct. There are many reasons for this:

- Human building activities destroy habitats.
- Humans often overhunt animals for food and pleasure.
- Human activities often pollute the environment.
- Humans outcompete animals for resources such as food.
- Climate change can cause species to die out, and human activities may contribute to this climate change.

Endangered species

Many animals and plants are not yet extinct, but their numbers are very low. They are known as **endangered species**, eg the giant panda and the Bengal tiger.

Endangered species will become extinct if:

- their numbers fall below a critical level
- the populations are restricted to small areas.
- there is not enough genetic variation in the population

- so the population cannot cope with environmental changes.

Conserving wildlife

Biologists now try to **conserve** endangered species and habitats using a number of approaches (see left). It is important to conserve wildlife for several reasons:

- Species may be useful in the future in the development of new medicines.
- Many species could represent future food supplies.
- Humans should consider damage not just to individual species but to food chains as a whole.
- Humans have a duty to conserve the environment for future generations to enjoy.

Conserving tigers

The world tiger population is now less than 3200 individuals, a figure that has dropped significantly due to hunting (for skins and medicines) and habitat destruction.

Previous conservation efforts have included a CITES ban on trade in tiger parts, which reduced hunting of tigers.

The World Wide Fund for Nature (WWF) launched a campaign in 2010 (the Chinese year of the Tiger) to double tiger numbers by 2022. The campaign aims to:
- Prevent the tiger population declining to a level where there is too little genetic variation for tigers to survive.
- Monitor the growth of tiger populations.
- Monitor the growth of prey populations as a food source.
- Create tiger conservation landscapes (TCL) as protected habitats.

Each of these aims must be checked regularly to evaluate the success of the programme.

Sustainable development

Humans need to manage the environment in a sustainable way. Two examples of sustainable management are sustainable fishing and sustainable forestry. Humans need resources like fish and wood, so strategies are used to maintain them.
- Education – to avoid overuse and increase recycling.
- Quotas – fishing limits used to prevent overfishing.
- Replanting – replanting woodland after felling.

Conserving whales

The different species of whale live in different areas of the oceans, depending on their food source. Some species are endangered and are close to extinction due to hunting.

commercial value
Whales are of commercial value to humans through tourism (when alive) and to produce food, oil, and cosmetics (when killed).

captive programmes
Keeping whales in captivity allows breeding programmes (increasing whale numbers), whale biology research, and tourism opportunities, but restricts freedom.

whale biology
Whale conservation is crucial, and will allow us to fully understand their communication systems, patterns of migration, and ability to survive at great depth.

whaling
Most countries have banned whaling in order to conserve the whale population. However, it can be difficult to obtain and police international whaling agreements, and some culling is permitted for research.

Sustainable development strategies are planned at local, national, and international level to be effective. They protect the population size of many species and have prevented species such as cod becoming endangered. The global human population is growing. To maintain sustainable development we must produce and use food and energy efficiently, and manage waste production carefully.

A Whales were once common in the oceans. Suggest a reason why they are now much more rare.

B Suggest advantages of keeping some species of whale in captivity.

Questions

1 Name two endangered and two extinct species. ↓E

2 Explain how fishing quotas should help maintain the cod stocks in the North Sea. ↓C

3 There is an international agreement which controls trade in endangered species. Suggest how this could help conserve species such as elephants. ↓A*

Module summary

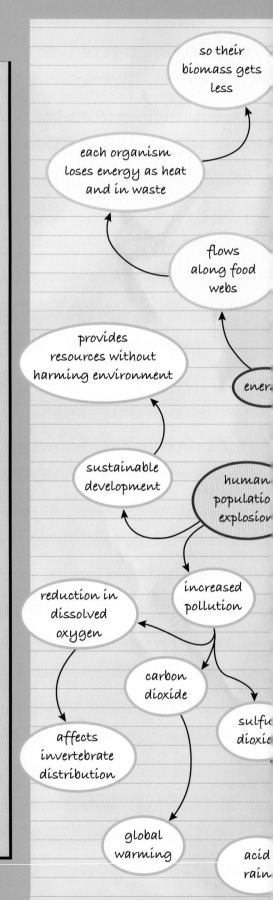

Revision checklist

- To make living things easier to name and study, we classify them into groups and subgroups.
- Organisms that interbreed and produce fertile offspring are members of the same species.
- Food webs show how energy from the Sun flows through ecosystems, from producers to secondary consumers.
- The mass of living material (biomass) gets less as you move along food chains.
- Energy flows from the plants that capture it to carnivores, but some is lost at each step as heat and in waste materials.
- The carbon in living things is recycled. Photosynthesis traps carbon dioxide in biomass. It passes along food chains, and is released when living things respire or decompose.
- Nitrogen is also recycled. Plants use nitrates from soil to build proteins, which pass along food chains. Decomposers return the nitrogen to the soil as ammonia, and bacteria convert this back to nitrates.
- All living things compete for resources. Plants compete for light, space, water, minerals, and carbon dioxide and animals for food, territory, mates, and water.
- Predators and prey have adaptations which aid their survival. Changes to one population have a direct effect on the other.
- Arctic animals are adapted to minimise heat loss, and desert animals are adapted to minimise water loss.
- Plants are adapted to control water loss and light absorption.
- Darwin's theory of evolution by natural selection explains how organisms change over time to suit changing habitats.
- Lamarck thought changes acquired during an organism's lifetime were passed on to offspring, but this theory has been discredited.
- The human population explosion has led to food shortages, increased pollution, and pressure on land and resources.
- Pollution causes acid rain and global warming. Indicator species show how polluted air or water is.
- Sustainable development allows us to get the resources we need without harming the environment.

NOW USE THE B2 GRADE CHECKER ON PAGE 242

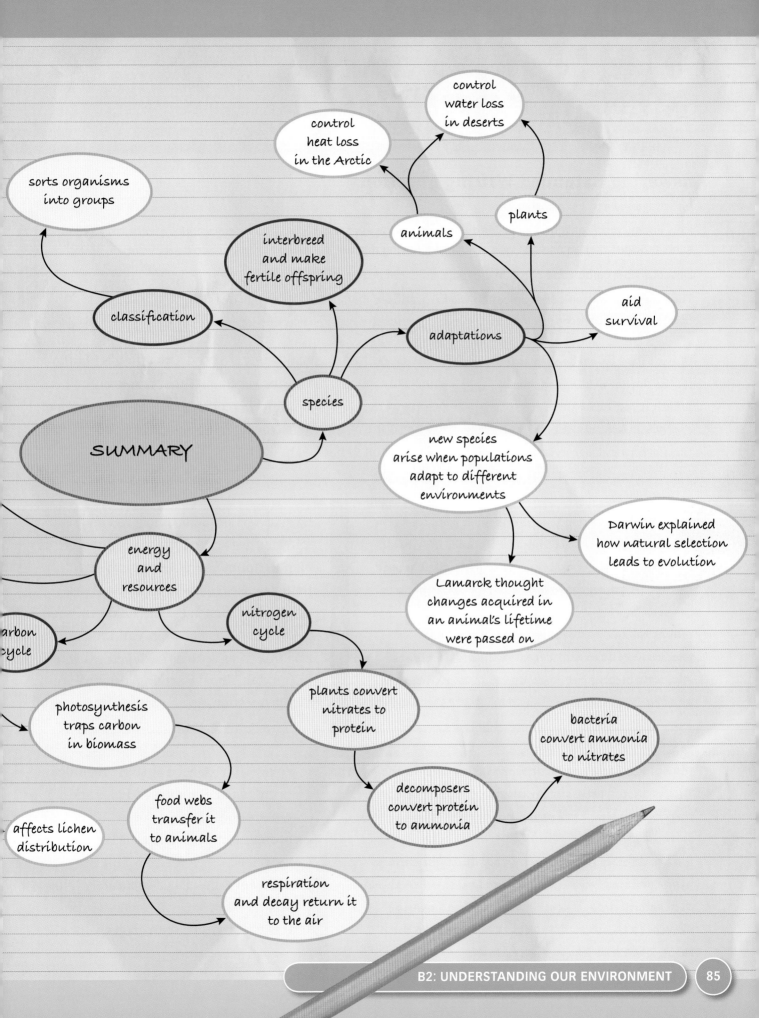

sorts organisms into groups

control heat loss in the Arctic

control water loss in deserts

classification

interbreed and make fertile offspring

plants

animals

adaptations

aid survival

species

SUMMARY

new species arise when populations adapt to different environments

Darwin explained how natural selection leads to evolution

energy and resources

nitrogen cycle

Lamarck thought changes acquired in an animal's lifetime were passed on

carbon cycle

plants convert nitrates to protein

bacteria convert ammonia to nitrates

photosynthesis traps carbon in biomass

decomposers convert protein to ammonia

affects lichen distribution

food webs transfer it to animals

respiration and decay return it to the air

Answering Extended Writing questions

QUESTION

New energy from sunlight reaches Earth every day. However, there is only a certain amount of other resources, such as water, carbon and nitrogen, available for living things. These resources have to be recycled.

Describe how carbon and nitrogen are recycled in nature.

The quality of written communication will be assessed in your answer to this question.

Plants take in carbon dioxide in the day to make food. They breath it out at night. Annimmals eat plants and then they breath out carbon dioxide.

↓ E

Examiner: No technical terms are used. Breathing is not the same as respiration. The answer implies that plants only respire at night, which is not correct. There is no description of the recycling of nitrogen. One mark for the reference to plants using carbon dioxide to make food. There are spelling errors.

Plants take in nitrogen from soil. It goes up to the leaves and they use it to make protenes. Animals eat plants and get protene. When they die bacteria break them down and their nitrogen goes back to the soil.
Plants take in carbon dioxide from air. They photosynthesis to make food. Animals eat plants. Animals breathe out carbon dioxide. Burning trees adds carbon dioxide to air and causes global warming.

↓ C

Examiner: Could be improved if 'nitrates' was used instead of 'nitrogen' as plants cannot use gaseous nitrogen. Better to use technical terms, such as 'decompose' instead of 'break down', and 'respiration' instead of 'breathing out'. Burning trees returns carbon dioxide to the air, but global warming is not relevant to this question. Some spelling mistakes.

Plants use carbon dioxide from the air and make food. This is photosynthesis. Animals get the carbon when they eat plants. Respiration puts carbon dioxide back in the air.
Plants take up nitrates from soil. They use them to make proteins. Animals eat the plants. When things die bacteria decay them. In soil, bacteria change ammonia to nitrates. Plants take these into their roots. DNA also has nitrogen in it.

↓ A*

Examiner: A good answer. Technical terms, such as photosynthesis and respiration, are used correctly. The recycling of carbon is clearly described. Then the recycling of nitrogen is described separately. This answer shows a good understanding. It is well written, with good spelling.

Exam-style questions

1 Match these words and meanings:

A01

Word	Meaning
consumer	an organism that makes its own food
the Sun	a step in the food chain
producer	energy source
trophic level	a herbivore or carnivore

2 Jen measured oxygen concentrations in a stream that runs past a pig farm.

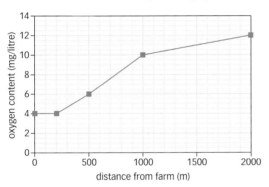

A02 **a** Where was the oxygen content of the water lowest?

A02 **b** How much did it rise between 500 m and 1000 m from the farm?

Jen took water samples and counted the number of mayfly nymphs in them.

Distance from farm (m)	Mayfly nymphs
0	0
200	0
500	6
1000	34
2000	35

A02 **c** What can Jen conclude about mayfly nymphs?

A02 **d** Jen has a hypothesis. She thinks that the pig farm is polluting the stream. Suggest other measurements she could take to strengthen her evidence.

3 Pyramids of biomass model the way the mass of living things drops as you move along a food chain.

A01 **a** Suggest two reasons why accurate pyramids of biomass are difficult to construct.

A01 **b** Explain why the biomass of carnivores is always less than the biomass of herbivores in a food chain.

A02 **c** In the oceans, food chains often have more than three steps. Suggest why.

E ↓

C ↓

A* ↓

Extended Writing

4 Kasia is a vegetarian. She says if

A02 everyone ate cereals instead of meat we could feed more people. Explain why she is right. 5 Plants absorb carbon dioxide during photosynthesis. Explain

A01 how carbon is returned to the atmosphere. Why is the amount of carbon dioxide in the atmosphere rising?

6 Lamarck and Darwin both proposed

A01 &2 theories that explained evolution. Using giraffes as an example, explain the similarities and differences between their theories.

C ↓

A* ↓

A01 Recall the science
A02 Apply your knowledge
A03 Evaluate and analyse the evidence

B2: UNDERSTANDING OUR ENVIRONMENT 87

B3
Living and growing

Why study this module?

You learn things in school to help you understand the world around you. Some of the main areas of research in biology today are genes, ageing, and regenerative medicine. In this module, you will learn how the genetic code governs the making of proteins in your cells, and why proteins are important. Enzymes are proteins, as is haemoglobin. You will find out why you need energy (for example, for your cells to make proteins), and how your cells respire to release energy from the food you eat. Your blood brings the glucose and oxygen for respiration to your cells.

You will also learn about selective breeding, genetic engineering, gene therapy for some inherited diseases, cloning, and about genetically modified crop plants.

You should remember

1   You are made of cells that are organised into tissues, organs, and systems – such as the reproductive system.

2   Plant and animal cells both have a membrane, cytoplasm, nucleus, mitochondria, and ribosomes, but plant cells also have a cell wall and a large vacuole.

3   There are two types of cell division: mitosis for growth and asexual reproduction; meiosis for sexual reproduction.

4   Cells need energy for their chemical reactions (metabolism) and for division.

5   Respiration releases energy from the food you eat.

6   Your blood is a transport system.

7   Genes, on chromosomes, determine your characteristics.

8   The cell that you developed from was made from your mother's egg and your father's sperm, and each contained your parents' genes.

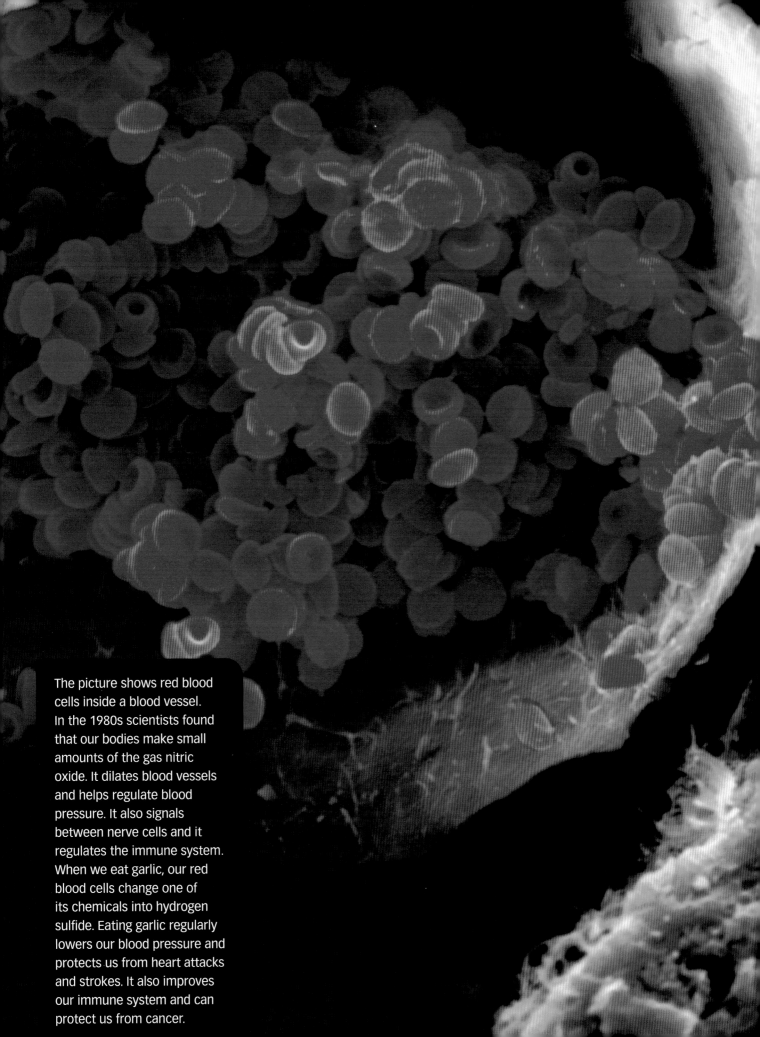

The picture shows red blood
cells inside a blood vessel.
In the 1980s scientists found
that our bodies make small
amounts of the gas nitric
oxide. It dilates blood vessels
and helps regulate blood
pressure. It also signals
between nerve cells and it
regulates the immune system.
When we eat garlic, our red
blood cells change one of
its chemicals into hydrogen
sulfide. Eating garlic regularly
lowers our blood pressure and
protects us from heart attacks
and strokes. It also improves
our immune system and can
protect us from cancer.

Learning objectives

After studying this topic, you should be able to:

✔ describe the structure and functions of a cheek cell

✔ know what chromosomes are and where they are found

✔ know that DNA controls the production of different proteins

Key words

gene, genetic code, base pairs, proteins

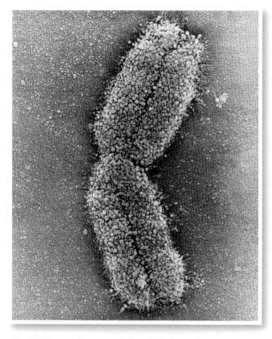

▲ DNA condenses and coils into a chromosome like this one (× 12000). This chromosome consists of two identical molecules of DNA, each one containing exactly the same alleles as the other. They are joined at a region near the middle. This gives the classic shape of visible chromosomes. It is only when chromosomes are coiled up like this that they take up stains and you can see them under a light microscope.

The structure of a cheek cell

Cells are the building blocks of living organisms. Cheek cells are typical animal cells. Each is enclosed in a membrane, contains cytoplasm and organelles such as mitochondria, and has a nucleus. Inside the nucleus is the DNA that carries the organism's genetic code.

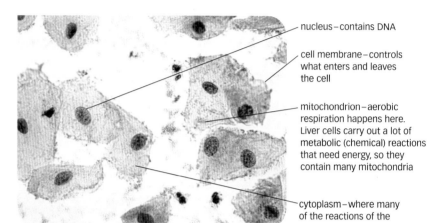

nucleus – contains DNA

cell membrane – controls what enters and leaves the cell

mitochondrion – aerobic respiration happens here. Liver cells carry out a lot of metabolic (chemical) reactions that need energy, so they contain many mitochondria

cytoplasm – where many of the reactions of the cell happen

▲ Light micrograph of cells that line the inside of your cheek (× 2600). The cells have been stained so that the normally transparent, colourless cytoplasm and the nuclei and other organelles can be seen. Respiration takes place in the mitochondria. At this magnification the ribosomes are too small to be seen clearly on most light microscopes. Ribosomes are where proteins are assembled from amino acids.

Chromosomes are made of DNA and contain genes

Inside the nucleus of each of your body cells you have 23 pairs of chromosomes. Each chromosome is one long coiled molecule of DNA. Within each DNA molecule, there are shorter sections of DNA. These sections are different **genes**.

Each gene has coded genetic information. This **genetic code** is formed by the sequence of **base pairs** in a particular length of DNA. Each gene contains a different sequence of base pairs. A gene codes for a particular combination of amino acids that makes a specific (particular) **protein**.

You need proteins for
- growth, which involves making new cells
- repair of damaged tissue, by replacing dead cells with new cells
- building structures such as muscle, bone, skin, hair, enzymes, hormones, antibodies, and haemoglobin.

Your genetic code controls how enzymes are made in your cells. Enzymes control all the chemical reactions that go on in your cells. So the genetic code controls all cell activity. As a consequence it controls most of your characteristics.

> Different types of cells have different functions and therefore need different proteins, including enzymes. Hence in any cell only some of the full set of genes are used, and the rest are switched off. The genes in a cell that are switched on determine the function of the cell.

Ideas about science: how the structure of DNA was worked out

In 1953 two scientists, Watson and Crick, at Cambridge University in the UK, used data from the work of other scientists to work out the structure of DNA.

Rosalind Franklin, at King's College in London, obtained X-ray data which showed that a DNA molecule consisted of two chains wound in a double helix. Watson and Crick used her data and help from chemist colleagues to build their model of DNA – a double helix with pairs of bases forming cross-links. They published their paper before Franklin and so they got all the credit. In 1962 they shared the Nobel Prize with Maurice Wilkins, Franklin's supervisor. Franklin died aged 37 in 1958 and the Nobel Prize is not given posthumously (after someone's death). In 1968 James Watson wrote a book and acknowledged the importance of Franklin's contribution.

> **Questions**
>
> 1 State two reasons why you need to make new proteins in your cells.
>
> 2 How does DNA carry the genetic code?
>
> 3 Explain how the genetic code controls cell activity.
>
> 4 Explain how Watson and Crick used data from the work of other scientists to build their model of DNA.
>
> 5 Explain why there is always a time delay (often of ten years or even more) between a scientific discovery and any reward, such as the Nobel Prize.

> A What is the function of: (a) mitochondria (b) the nucleus (c) the cell membrane (d) the cytoplasm?
>
> B What are chromosomes made of?
>
> C Where are your genes?

There is always a delay between a discovery and its importance being recognised or rewarded. Other scientists have to repeat the work to verify it. And it is not always clear, straight away, how important and useful the discovery is.

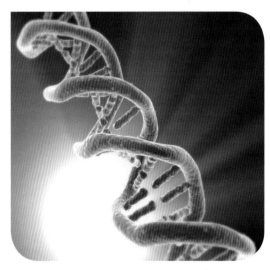

▲ Model of a section of DNA, showing the double helix and the cross-links formed by pairs of bases

Exam tip OCR

✔ Try and refer to the bases in DNA as 'DNA bases'. The word 'base' has another meaning – in chemistry, alkaline substances are called bases.

Key words

complementary base pairing, **base triplet**, **ribosomes**, **mRNA**

▲ You can see how the bases pair up in the DNA molecule. Green = T; red = A; blue = G; yellow = C.

Exam tip

✔ You must learn that A pairs with T and C pairs with G.

Genes and bases

Genes are in chromosomes so they cannot leave the nucleus. Copies of a gene pass out of the nucleus and go to the ribosomes, where proteins are assembled. Each gene (section of DNA) has a sequence of DNA bases in it. There are four different bases.

The four bases are called A, T, G, and C. You do not need to know their names. These bases form the cross-links in the DNA molecule.

A always pairs with T. G always pairs with C. This is known as **complementary base pairing**.

- The base pairs form a code.
- They are 'read' in groups of three, or **base triplets**.
- Each triplet specifies a particular amino acid. So ATC will specify a different amino acid from ACT.
- The sequence of the triplets of bases on a section of DNA specifies the sequence of amino acids in a protein.
- As you know, DNA always stays in the nucleus, and proteins are assembled at **ribosomes**, in the cytoplasm of the cell.
- So another molecule, called **mRNA**, carries a copy of the coded instructions in a gene out of the nucleus. The instructions in the DNA are like a recipe that is in a book you cannot take out of the library. The mRNA is like a photocopy of the recipe that you can take out of the library.
- The mRNA is a single-stranded molecule. It is a copy of one strand of a length of DNA. In other words, it is a copy of a gene.
- The mRNA goes to a ribosome. Here it governs how the amino acids are assembled into a protein.
- The amino acids, which you get from eating and digesting protein in food, are assembled into long chains.
- The sequence of amino acids in the protein governs how the protein will fold up into a particular shape.

Each different type of protein has a specific shape. This is how enzymes each fit just their own specific substrate molecule. You can read more about this on spread B3.4.

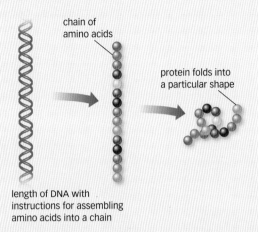

chain of amino acids

protein folds into a particular shape

length of DNA with instructions for assembling amino acids into a chain

▲ Simplified diagram to show how the coded information in a gene determines the shape and the function of a protein.

A Where, in the cell, are proteins made?

B Where have the amino acids, which are made into proteins in your cells, come from?

C Explain how the sequence of bases in a length (section) of DNA determines the sequence of amino acids in a protein.

How proteins determine your characteristics

All proteins have a specific shape, and this enables them to carry out their function. The characteristics that you inherit involve proteins. These characteristics may rely on the help of enzymes and hormones to develop.

Your cells will not function without proteins, such as enzymes.

Questions

1 Explain the role of mRNA in making proteins.

2 What is complementary base pairing?

3 What determines how a protein will fold up into a particular shape?

4 Why is the shape of a protein molecule very important?

5 Explain how genes control cell function.

A*

Learning objectives

After studying this topic, you should be able to:

- ✔ recall examples of proteins including collagen, insulin, and haemoglobin
- ✔ know that only some of the genes in a cell are used

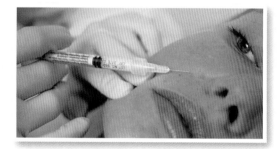

▲ Collagen injections can smooth out wrinkles in the skin

Key words

structural protein, hormone, carrier protein

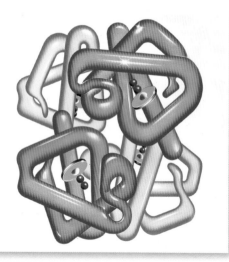

▲ The structure of haemoglobin. The pink and blue areas are proteins. The green areas are where the iron is. Each iron atom can hold two oxygen atoms – shown as red spheres.

Collagen

Your skin, bones, cartilage, tendons, ligaments, and walls of blood vessels contain a type of protein called collagen.

▲ Collagen fibres, seen using an electron microscope

Because collagen makes up some of the structure of your body, it is called a **structural protein**.

Insulin

Insulin is a protein **hormone** made in the pancreas. It travels in the blood stream to the target organs, your muscles, and your liver. It regulates your blood sugar level. Many other hormones are also proteins.

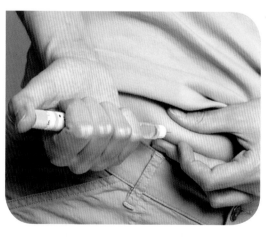

◀ People with type 1 diabetes have to inject themselves with insulin. Because insulin is a protein, it would be digested if they took it in by mouth.

Haemoglobin

You have haemoglobin in your red blood cells. Haemoglobin carries oxygen from your lungs to your respiring cells.

Haemoglobin is a **carrier protein**.

Enzymes

Enzymes are catalysts that control all the chemical reactions in your cells and in your digestive tract (gut). You will learn more about them on spread B3.4.

In addition, you also have many other proteins that do important jobs. These include:

- antibodies
- receptors for hormones on membranes of target cells
- channels in cell membranes.

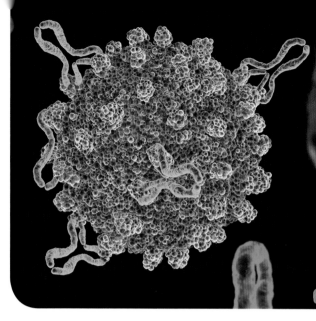

▲ Antibodies surrounding a virus particle in the blood. Antibodies are proteins. Each type can fit onto the particular antigens, also made of protein, on the particular virus coat.

> **A** Why is collagen called a structural protein?
>
> **B** What is the function of the protein insulin?
>
> **C** Why is haemoglobin described as a carrier protein?
>
> **D** Name four other types of proteins in your body, besides the three mentioned above.

Different types of cell make different proteins

You have about 220 different types of cell in your body. However, you began life as one cell – a fertilised egg, called a zygote. This cell is the ultimate stem cell. It can divide to make more cells and these can become any one of the 220 different cell types.

In any cell, not all of the 20 000 genes are switched on and being used.

Different cell types have different jobs to do. Therefore they need different enzymes to catalyse the particular chemical reactions that only they carry out. Only they need the genes that code for those enzymes to be switched on.

Particular cells also need to make other specific types of protein. Cells in the pancreas need to make insulin, but liver or muscle cells do not need to make insulin. Liver and muscle cells need to make receptors on their membrane that fit insulin molecules. Red blood cells are the only cells that need to make haemoglobin.

Some genes do not actually control making proteins. They switch other genes on or off.

Exam tip OCR

✔ Remember that a red blood cell does not have a nucleus. So if you are asked how many chromosomes it has, the answer is 0.

Questions

1 Name one protein made in red blood cells that is not made in white blood cells.

2 Name a protein that is only made in cells of the pancreas.

3 Name a protein that is only made in cells of the stomach.

4 Explain how switching genes on or off can lead to cells becoming specialised for their different functions.

Learning objectives

After studying this topic, you should be able to recall that enzymes:

- ✔ are proteins that catalyse chemical reactions in living cells
- ✔ have high specificity for their substrates
- ✔ work best at particular temperatures and pH

A What is a catalyst?

B Name three types of chemical reaction that enzymes speed up in living organisms.

Key words

enzyme, catalyst, substrate, specific, optimum, denatured

Enzymes are biological catalysts

Enzymes are biological **catalysts** because they speed up chemical reactions.

Most of these reactions, such as
- photosynthesis
- respiration
- protein synthesis

take place inside living cells.

Enzymes can be used to catalyse the same type of reaction many times. This is like using one type of screwdriver to screw in many of the same type of screw, one at a time.

The shape of an enzyme is vital for its function

Enzymes, like all proteins, are folded into a particular shape. The shape of one particular area of the enzyme molecule, called the active site, is very important.

- The **substrate** molecules fit into the active site.
- This brings them together so they can form a bond.
- This makes a bigger molecule.

In some cases (as shown on the left):
- a big substrate molecule fits into the active site
- a bond breaks
- two smaller product molecules are made.

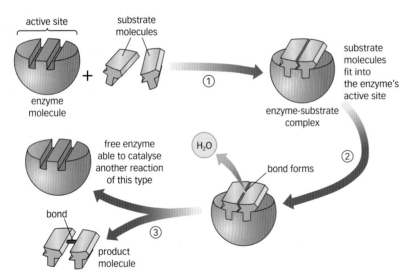

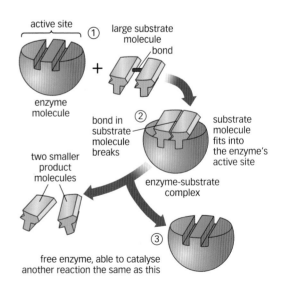

▲ The lock and key hypothesis is a hypothesis about how enzymes work. The two substrate molecules fit side by side into the enzyme's active site. A bond forms between them and one large product molecule is formed.

▲ The large substrate molecule fits into the enzyme's active site. A bond breaks and two product molecules are formed.

Enzymes have high specificity for their substrate

Only one particular type of substrate molecule can fit into an enzyme's active site. This is like the way only one type of key will fit into a particular lock. This means each enzyme is **specific** for its substrate molecules.

What makes enzymes work best?

Each enzyme works best at a particular temperature, known as its **optimum** (best) temperature, and at its optimum pH.

Low temperatures

The enzyme and substrate molecules have less energy. They do not move very fast so they do not collide (bump into each other) very often. The rate of reaction is low.

High temperatures

As the temperature increases, the enzyme and substrate molecules move more quickly and collide more often. This gives a faster rate of reaction. For most chemical reactions, the rate of reaction doubles with a 10 °C rise in temperature.

However, if the temperature becomes too high then:
- The shape of the enzyme's active site changes.
- The substrate molecule cannot fit into the active site.
- The rate of reaction slows and eventually stops.

When the shape of the enzyme has changed in this way, it cannot go back to its original shape. The change is irreversible. The enzyme is **denatured**. Because the shape of the enzyme's active site is altered at high temperatures, rate of reaction does not always double with every 10 °C temperature increase. Q_{10} (the temperature coefficient) can be calculated using the formula:

$$Q_{10} = \frac{\text{rate at higher temperature}}{\text{rate at lower temperature}}$$

pH

Each type of enzyme works at an optimum pH. If the pH changes very much then:
- The shape of the active site changes.
- The substrate molecules cannot fit into it.
- The enzyme has been denatured.

C What are enzymes made of?

D Why is the shape of the active site of an enzyme important?

E How is each enzyme specific for a particular substrate?

Did you know...?

Many enzymes in the body could work more quickly at temperatures above 37 °C. However, if we kept our bodies hotter than this, many of our other proteins would be damaged. At 37 °C our chemical reactions go on fast enough to sustain life. But some bacteria can live in very hot places, and their enzymes work well at 100 °C.

Questions

1 State two conditions that enzymes need to work best. E

2 Describe how increasing the temperature from 10 °C to 25 °C makes the rate of a reaction increase.

3 Describe how if the temperature increased to 60 °C, the rate of the reaction would slow down and eventually stop. C

4 As well as having enzymes inside your cells, you also have them in your blood. Your blood pH needs to be kept very close to 7.2. Why do you think this is? A*

Learning objectives

After studying this topic, you should be able to:

- ✔ explain that mutations may lead to the production of different proteins
- ✔ know that different organisms produce different proteins

A What is a gene mutation?

B When is a gene mutation likely to occur spontaneously in a cell?

C Why may changing the sequence of amino acids in a protein stop the protein from functioning or change its function?

Did you know...?

Scientists who study evolution compare the proteins, and the genes that code for them, of different species of organism. The more these proteins and genes are similar, the more closely related the different species are.

Gene mutations

A **gene mutation** is a change to a gene. Mutations occur **spontaneously**, with no external cause. Often it is a copying error when DNA is replicating itself. Mutations can also be caused by

- chemicals, such as tar in tobacco smoke
- **ionising radiation**, such as X-rays and ultraviolet light.

This involves changing a base pair in the section of DNA. The sequence of its bases is changed. This changed gene may not now code for the protein it normally codes for. Instead, it may now code for a different protein, which has a different shape and may not be able to do its normal job. For example, if the shape of an enzyme's active site changes, it cannot join with its substrate molecule. It cannot catalyse the chemical reaction. However, it may be able to catalyse a different reaction.

Some mutations are harmful

Mutations may cause

- cells to keep on dividing (cancer)
- a particular enzyme not to work, causing a serious illness
- slightly differently shaped haemoglobin molecules (sickle cell anaemia).

A gene mutation that causes a protein channel, in cell membranes lining the airways, to be different and not function causes the genetic disease cystic fibrosis.

▲ These cancer cells have undergone mutations. They keep on dividing and they have lost their normal shape.

Key words

gene mutation, spontaneous, ionising radiation

Some mutations can be useful

Pale skin in humans is caused by mutations to the genes that control skin colour. It is not useful to people who live in hot regions, because the skin has less protection against the strong sunlight. However, it is useful to people who live in temperate regions of the world. It allows their skin to make vitamin D because the weaker sunlight can penetrate their paler skin. Early humans with this mutation could live in temperate regions without getting rickets, which can be fatal.

▲ A range of skin colours

Some mutations are neutral

Being unable to roll your tongue is caused by a gene mutation. However, it does not appear to cause anyone any harm.

Nor does it seem to be useful. Some of us have free ear lobes and some of us have attached lobes. Neither seems either useful or harmful.

Mutations lead to different proteins being made

Proteins are made of long chains of amino acids. Genes are the instructions for assembling amino acids into a protein.

> Each protein has its own number and sequence of amino acids. As a result, each protein then folds into a particular shape to carry out its function.

All species on Earth have evolved from older ancestral organisms. Mutations have caused differences between the genes of different species. As a result, different species make different proteins because their genes have slightly different instructions for the assembly of amino acids. If the amino acid sequence in a protein is changed then that protein folds in a different way and may perform a different function.

Questions

1 Name three diseases caused by gene mutations. ↓ E

2 Explain why the mutation to produce pale skin in humans can be a useful mutation.

3 Rabbit fur is normally a grey/brown colour, called agouti. A mutation can lead to a white coat. In what circumstances would a white coat be useful, and why? When do you think having a white coat would be harmful for the rabbit? ↓ C

4 Can you think of another gene mutation in humans that is neutral?

5 The Inuit people live in temperate/cold regions. However, they have quite dark skin. Unlike all other humans, they do not need to make vitamin D by the action of sunlight on the skin. Why do you think this is? ↓ A*

▲ An Inuit hunter in Greenland

Exam tip OCR

✔ For questions like Question 5 above, you need to apply your knowledge to a new situation to work out the answers. Don't be put off by the subject matter.

▲ Buffalo (*Bison bison*) grazing on grass

A State three reasons why living things need energy.

B Name three types of large molecules that are made in living cells using energy.

▲ This grey wolf (*Canis lupus*) needs energy to run

What is energy?

Energy is the ability to do work. All matter has energy. There are different forms, such as kinetic (movement), potential (stored), heat, sound, electrical, and light energy. Each form of energy can be transformed into another form.

- Plants trap sunlight energy and use it to make large molecules – proteins, fats, and carbohydrates. These molecules contain stored energy.
- Animals get these molecules, containing stored energy, by eating plants or eating other animals that have eaten plants.

Why do living organisms need energy?

All life processes in all living organisms (including plants as well as animals) need energy. The energy may be used

- to build large molecules from smaller ones
- for muscle contraction in animals
- to control body temperature in mammals and birds.

Building large molecules from smaller ones

- Plants use sugars, nitrates, and other nutrients to make amino acids.
- Amino acids are joined together in long chains during a process called **protein synthesis**. All living things need to make proteins such as enzymes and parts of their structure.
- Plants join sugar molecules together to make starch.
- Animals join sugar molecules together to make glycogen, which is similar to starch.
- Living organisms join fatty acids and glycerol together to make lipids (fats).

Muscle contraction

Animals need to move, to find food or a mate, or to escape from predators. Muscle contraction needs energy and causes movement.

Controlling body temperature

Some organisms cannot control their temperature very well. As the surrounding temperature changes, so their temperature may change. They control it by moving into the shade or into a warmer place. Snakes and lizards are very slow and sluggish in winter, or at night, when it is cold.

Birds and mammals can be active at night and during the winter. This is because a lot of the energy from the food they eat is released as heat energy. This keeps their body temperature steady regardless of the external temperature. However, it means that they need to eat more food than animals such as fish, snakes, and lizards.

How is energy released from food molecules?

Respiration in living cells releases energy from glucose molecules. You get glucose when you digest carbohydrates that you eat. Respiration is a process that involves many chemical reactions, all controlled by particular enzymes.

Aerobic respiration

Aerobic respiration uses oxygen. It happens continuously in the cells of plants and animals.

Anaerobic respiration

Anaerobic respiration is a different type of respiration, that takes place without oxygen. This does *not* happen continuously in plant and animal cells. It happens when cells are not getting enough oxygen.

Questions

1 Explain why animals need energy for movement.
2 What process in cells releases energy from food?
3 Explain why birds and mammals can be active at night when it is cold.
   ↓ E
4 Explain the difference between aerobic and anaerobic respiration.
   ↓ C
5 On a cold night in winter, a robin will lose a quarter of its body mass. Why do you think this is?
6 During the winter in the UK many birds, such as swallows, cannot find enough food to eat to keep warm. How do you think they solve the problem?
   ↓ A*
7 During the winter in the UK some mammals, such as hedgehogs, cannot find enough food to eat to keep warm. How do you think they solve the problem?

ATP

The energy released from glucose during respiration is used to make molecules of ATP (adenosine triphosphate). ATP is used as the energy source for all processes in cells that need energy.

Did you know...?

The average temperatures of different species of birds and mammals vary.

Average body temperatures of some mammals and birds

Animal	Average body temperature (°C)
human	37.0
chimpanzee	37.0
dog	38.0
cat	39.0
rabbit	39.5
chicken	42.0
owl	38.5
eagle	48.0
penguin	38.0

Key words

protein synthesis, respiration, aerobic, anaerobic

Key words

lactic acid, **muscle fatigue**, **oxygen debt**

Exam tip   OCR

✔ Remember that aerobic means with oxygen. The prefix 'an' or 'a' means without. So anaerobic means without oxygen.

Metabolic rate

Because aerobic respiration uses oxygen, the amount of oxygen an organism uses in a particular time period (its rate of oxygen consumption) indicates its metabolic rate. The metabolic rate is a measure of how quickly all the chemical reactions are going on in the organism's body. How much carbon dioxide it produces in a minute also indicates an organism's rate of respiration.

Respiration at rest

You have learnt that respiration provides the energy needed for all life processes. Living organisms respire all the time. Aerobic respiration uses oxygen. During aerobic respiration in living cells, there are chemical reactions that

• use glucose sugar and oxygen
• and release energy.

Aerobic respiration involves a series of several chemical reactions. However, the whole process can be summarised simply by the following equation:

$$\text{glucose} + \text{oxygen} \rightarrow \text{carbon dioxide} + \text{water} \,(+\, \text{energy})$$

$$\underset{\substack{\text{(contains stored}\\ \text{energy)}}}{C_6H_{12}O_6} + 6O_2 \rightarrow 6CO_2 + 6H_2O \,(+\, \text{energy})$$

Aerobic respiration takes place in mitochondria. Liver cells carry out many reactions, so they need lots of ATP (energy) and have lots of mitochondria to supply this. Muscle cells also have lots of mitochondria, as they require a lot of ATP for contraction.

Respiratory quotient (RQ)

This is calculated using the formula:

$$\text{respiratory quotient} = \frac{\text{volume } CO_2 \text{ produced}}{\text{volume } O_2 \text{ consumed}}$$

So if 10 cm³ carbon dioxide is produced and 10 cm³ oxygen is used by respiring tissue in the same period, the RQ is 1.

Respiration during exercise

During hard exercise your muscles need more energy, so your rate of respiration increases. You will

• use more oxygen per minute
• produce more carbon dioxide per minute.

Your heart rate (and therefore your pulse rate) goes up to deliver more oxygen and glucose to your muscles per minute. Your breathing rate goes up to remove the extra carbon dioxide more quickly.

A Explain why your pulse rate increases when you run.

Anaerobic respiration

When you start hard exercise, your heart rate does not go up quickly enough to supply the extra oxygen. To make up for the shortfall in energy release, your muscle cells use anaerobic respiration as well as aerobic respiration.

Glucose is incompletely broken down to **lactic acid**.

glucose   →   lactic acid
(contains stored energy)   (+ energy)

Anaerobic respiration releases much less energy per molecule of glucose than aerobic respiration. However, this incomplete breakdown happens quickly. Many molecules of glucose can quickly be partly broken down. Anaerobic respiration cannot go on for long because the lactic acid is toxic, causing pain and fatigue. as it accumulates in muscle cells.

Enzymes and respiration

The reactions in respiration are controlled by enzymes. Rate of respiration is influenced by temperature and pH.

Temperature

When people warm up before doing strenuous activity, their muscles warm up a bit and the respiration reactions go more quickly. When they start exercising hard, their respiration is faster and can release more energy.

pH

The increased lactic acid from anaerobic respiration lowers the pH. This reduces enzyme activity, and so reduces the rate of respiration. Muscles get **fatigued**. This is painful and your muscles stop contracting.

Fatigue and oxygen debt

During hard exercise your muscles become fatigued due to lack of oxygen. They have to respire anaerobically as well as aerobically. Anaerobic respiration breaks glucose down incompletely and produces lactic acid. When you stop exercising
- your heart rate stays high so the blood can quickly carry lactic acid to the liver
- you continue to pant to breathe in extra oxygen to deal with the lactic acid in the liver.

The amount of oxygen needed to do this is called the **oxygen debt**.

Did you know...?

You can measure your pulse rate by placing your first and middle fingers on the radial pulse at your wrist and counting the number of pulses you feel in 15 seconds. Multiply this by four to calculate beats per minute. You could also use a pulse monitor on your index finger with a data logger.

Questions

1 Explain why your breathing rate increases when you cycle hard.

2 Why do your muscle cells use anaerobic respiration as well as aerobic respiration when you start to run fast?

3 Which provides more energy for each molecule of glucose used – anaerobic or aerobic respiration?

4 How will lactic acid affect the pH of your blood?

5 What causes muscle fatigue?

6 Why do you think an athlete's heart rate and breathing rate stay high for several minutes after running a 100 m sprint?

7 Why do you think snakes are slow and sluggish on a cold morning? Remember, if it is cold outside, the snake's body is also cold.

Learning objectives

After studying this topic, you should be able to:

- ✔ know that mitosis is a type of cell division in body cells producing two genetically identical cells
- ✔ know that organisms that reproduce asexually use mitosis

Key words

mitosis, chromosome, diploid, asexual reproduction, allele, growth

A Name a type of cell in your body that does not have any chromosomes in it.

Why do body cells divide?

Body cells divide

- to replace worn out cells
- to repair damaged tissues
- to grow by producing more cells.

Body cells divide by **mitosis**. Each cell produces two genetically identical daughter cells. This increases the total number of cells in a multicellular organism.

In the nucleus of most of your body cells you have two sets of **chromosomes**, arranged as 23 matching pairs of chromosomes. These cells are described as **diploid**, and are common to all mammals.

Copying the cell's genetic material

Before a cell divides, its genetic material has to be copied so that each cell has a complete set of genetic material. Each chromosome, made of one molecule of DNA (the genetic material), is copied. So before a cell divides, each molecule of DNA copies itself. This is called DNA replication.

How DNA replicates

- DNA is a double-stranded molecule.
- The molecule 'unzips', forming two new strands.
- This exposes the DNA bases on each strand.
- Spare DNA bases in the nucleus line up against each separated strand of DNA.
- They only align next to their complementary DNA base, forming base pairs.
- One molecule of DNA has become two identical molecules.

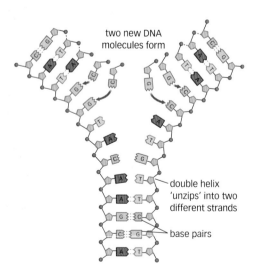

two new DNA molecules form

double helix 'unzips' into two different strands

base pairs

▲ How a DNA molecule replicates

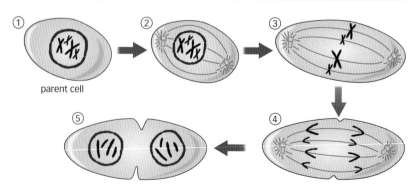

parent cell

▲ A cell dividing into two genetically identical cells by mitosis

How mitosis happens

- When each chromosome has made a copy of itself, these duplicated chromosomes line up across the centre of the cell.
- Then each 'double' chromosome splits into its two identical copies.
- Each copy moves to opposite ends of the cell.
- Two new nuclei form, each with a full set of chromosomes.
- The cell divides into two genetically identical cells.

Cells in the root tip of a hyacinth plant undergoing mitosis (×185)

Asexual reproduction

Some organisms can reproduce **asexually**. This type of reproduction uses mitosis. The cells produced by mitosis are genetically identical to the parent cell. They have the same **alleles** (versions of genes) as the parent.

Advantages of being multicellular

Early life forms on Earth were single-celled. There are still many simple single-celled organisms, such as the amoeba. Many organisms are now multicellular. Being multicellular means the organisms can be larger, can have different types of cells that do different types of jobs, and can be more complex.

However, being large and multicellular means that not all cells in a multicellular organism are in contact with the outside environment. This means they cannot rely on diffusion alone to receive the oxygen and nutrients they need and remove their waste. As a result they need organ systems such as specialised transport systems.

Mitosis in mature organisms

In mature animals cell division is mainly restricted to replacement of cells and repair of tissues. Mature animals do not continue to **grow**.

However, mature plants still have areas, such as root and shoot tips, where they can grow. The new cells made in these areas, by mitosis, can differentiate (become different and specialised) into many different types of plant cell.

Exam tip

✔ Mitosis replaces damaged cells and repairs tissue. It does not repair cells.

Specialised organ systems

Large multicellular organisms require specialised organ systems, such as:

- the nervous and endocrine systems, allowing communication between cells
- the circulatory system, supplying cells with nutrients
- the respiratory and digestive systems, controlling exchanges with the environment.

Questions

1 What are the advantages to an organism of being multicellular?
2 Explain why a cell's genetic material has to be copied before it divides by mitosis.
3 What is DNA replication?
4 Explain why mitosis is used for asexual division.
5 How is mitosis used in mature plants and animals?
6 Explain how mitosis happens.

Learning objectives

After studying this topic, you should be able to:

✔ know that gametes are made by meiosis for sexual reproduction

✔ know that gametes are haploid and combine to give a diploid zygote

✔ understand that meiosis produces genetic variation

A What are gametes?

B Where in the body are female gametes made? Where in the body are male gametes made?

Key words

gametes, meiosis, haploid, zygote, fertilisation

Exam tip | OCR

✔ Learn the spelling of meiosis. The word is similar to mitosis, so it has to be spelled correctly so that the examiner can be sure which type of cell division you are referring to.

Gametes

Gametes are sex cells. They are involved in sexual reproduction.

- Egg cells are made in the ovaries and sperm cells are made in the testes.
- Gametes are made by a special kind of cell division called **meiosis**.

How meiosis happens

- Just before the cell divides by meiosis, copies of the genetic information are made, just as they are before mitosis.
- So each chromosome has an exact copy of itself.
- However, in meiosis, the cell divides twice, forming four gametes.
- In the first division the chromosomes pair up in their matched pairs.
- They line up along the centre of the cell.
- The members of each pair split up and go to opposite poles (ends) of the cells.
- Now these two new cells each divide again.
- This time the double chromosomes split and go to opposite poles.
- Four cells, each genetically different from each other and from the parent cell, and with only half the number of chromosomes of the parent cell.

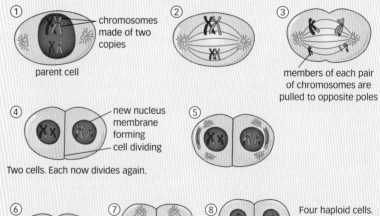

① chromosomes made of two copies

parent cell

② ③ members of each pair of chromosomes are pulled to opposite poles

④ new nucleus membrane forming — cell dividing ⑤

Two cells. Each now divides again.

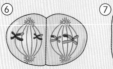

⑥ ⑦ ⑧ Four haploid cells. These are genetically different from each other and from the parent cell.

▲ A cell dividing into four haploid cells by meiosis

Meiosis introduces genetic variation. The cells made by meiosis are **haploid** gametes. They contain half the diploid number of chromosomes, ie they have just one set of chromosomes and not two. In humans, gametes have 23 chromosomes and are genetically different from the parent cell.

When two haploid gametes (an egg and a sperm) join, they produce a diploid cell called a **zygote**. This zygote will divide by mitosis into many cells and grow into a new individual.
- The joining of two gametes is called **fertilisation**.
- The combining of genetic material from two parents produces a unique individual.
- Half its chromosomes (and genes/alleles) have come from one parent and half from the other parent.
- It will have two sets of chromosomes.
- The combination of alleles will control the characteristics of the individual resulting from the zygote.

Gametes are adapted to their functions

Sperm cells (male gametes)
- are small and have a tail so they can swim to the egg cell
- have a nucleus to carry their genetic material
- are made in large numbers to increase the chance that one will find the egg
- have many mitochondria to provide a lot of energy
- have an acrosome that releases enzymes to digest (break down) the egg membrane.

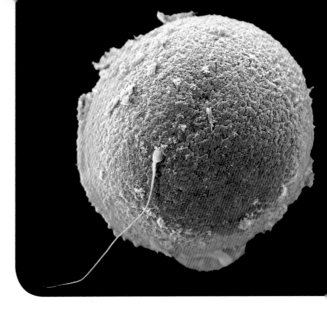

▲ Coloured electron micrograph showing a human sperm (coloured blue) penetrating a human egg (×400)

Did you know...?

- Sperms are released in a sugary fluid. Sperms have many mitochondria so they can release enough energy from the sugar to swim. Their mitochondria do not enter the egg. So all your mitochondria have come from the mitochondria that were in the egg – they have all come from your mother.
- Lots of sperm cells cluster around an egg. They each release enzymes to break down the tough egg membrane. This allows one sperm head to get into the egg. Then the egg membrane toughens up again so no more sperm cells can get in.
- Egg cells (female gametes)
    - are large and contain lots of stored food
    - have a nucleus to carry their genetic material
    - are produced in small numbers.

Questions

1 How are gametes different from body cells?

2 Describe how (a) male, and (b) female gametes are adapted to their function.

3 Why do sperm cells need a lot of energy?

↓ E

4 What do the following terms mean?
(a) fertilisation (b) haploid (c) diploid (d) zygote

5 What type of cell division do you think causes the zygote to develop into an embryo?

6 By what process will the mitochondria in sperm provide energy?

↓ C

7 Explain why sexual reproduction produces genetically unique new individuals.

↓ A*

Learning objectives

After studying this topic, you should be able to:

- ✔ recall the functions of blood cells
- ✔ recall the functions of plasma
- ✔ explain how the structure of a red blood cell is adapted to its function

Did you know...?

Five million is a big number. If you counted at the rate of 1 per second for 16 hours a day, it would take you 3 months to reach 5 million. If you have 5 million red blood cells in 1 mm³ of blood, then in your total blood volume of 5 litres, you have 5 million × 1 million × 5 = 25 million million ($25 × 10^{12}$) red blood cells.

Key words

haemoglobin, **oxyhaemoglobin**, **plasma**

A What is the function of red blood cells?

B Describe how the structure of a red blood cell enables it to carry out its function.

Blood platelets

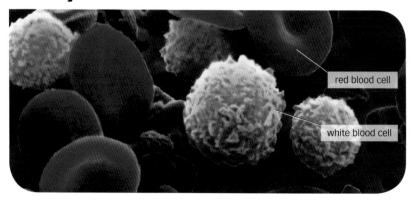

red blood cell

white blood cell

▲ Scanning electron micrograph of blood, with colour added

Red blood cells

Your red blood cells carry oxygen from your lungs to respiring cells in your tissues. There are many of these cells, about 5 million in 1 mm³ of blood (a drop of blood about the size of a pinhead). Each cell is well adapted to carry out its function.

Size: they are small in diameter and relatively thick. This means:

- They can just fit through capillaries, one at a time.

- Each has a large surface area compared to its volume (SA/V ratio) so a lot of oxygen can diffuse through the outer surface and into the centre of the cell.

Shape: they are biconcave discs (doughnut shaped). This increases the amount of surface compared to the inside volume even more.

No nucleus: before each red blood cell leaves the bone marrow, where it is made, and enters the blood, its nucleus breaks down. This leaves more room for lots of haemoglobin. (They also do not have other structures found in most cells, such as mitochondria or ribosomes.)

Haemoglobin: this is a carrier protein (see spread B3.3).

- In the lungs the haemoglobin in red blood cells reacts with oxygen, forming **oxyhaemoglobin**.
- At respiring tissues the oxyhaemoglobin breaks down to haemoglobin and oxygen. The oxygen is delivered to the respiring cells.

White blood cells

Your white blood cells help defend you against disease. Some ingest (eat) bacteria or viruses. Some make antibodies.

Platelets

Platelets are small cells that help your blood to clot when you cut yourself. This

- stops you bleeding
- prevents bacteria from entering the wound.

Plasma

Plasma is the straw-coloured liquid part of the blood. Over 90% of it is water.

Dissolved in it and being carried are

- digested food, such as amino acids and glucose, from the gut to body cells
- cholesterol, from the liver to cells
- hormones, from the glands where they are made to their target cells
- antibodies
- excess water from the gut, to the kidneys to be removed
- waste products such as carbon dioxide from respiring cells to the lungs, and lactic acid from muscles to the liver
- excess heat, from respiring cells to the skin.

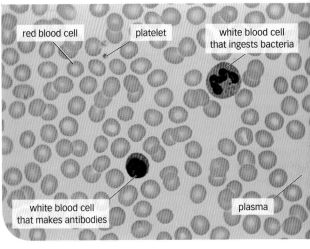

▲ Human blood seen under a light microscope (× 650). The white blood cells that make antibodies are about the same size as red blood cells.

Questions

1 What are the functions of: (a) white blood cells (b) platelets?

2 Explain how white blood cells carry out their function.

3 Red blood cells have no nucleus or other structures in the cytoplasm that most other cells have. State three things that you think they cannot do.

4 How do you think your blood plasma changes just after you have eaten a meal?

5 How do you think your blood plasma changes just after you have been running?

6 In your lungs, what happens to the haemoglobin in your red blood cells?

Exam tip OCR

✓ Always use the proper technical terms. Describe red blood cells as biconcave discs.

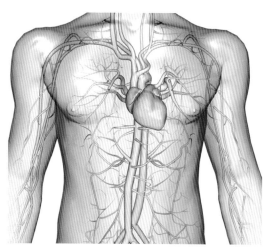

▲ The blood vessels of the body. Arteries are shown as red and veins as blue.

Exam tip   OCR

✔ Blood leaves the **v**entricles (lower chambers) of the heart in **a**rteries. It returns to **a**tria (top chambers) in **v**eins. So remember: it is always V and A.

How blood moves around your body

Like all mammals, your blood is in blood vessels. Your heart is a pump that creates enough force to make the blood circulate.

- Blood leaves the heart at high pressure in **arteries**.
- It returns from the body tissues to the heart in **veins**.
- Between the arteries and veins, at tissues, are **capillaries**. At capillaries glucose and oxygen leave the blood and enter cells, whilst waste carbon dioxide leaves the cells and enters capillaries.

Blood in your body moves from areas of high pressure to areas of low pressure. As blood moves through the circulatory system, pressure in the vessels is dropping. Veins therefore carry blood at the lowest pressure. Each type of blood vessel is adapted for its function.

Arteries

Arteries have a thick muscular and elastic wall. They can withstand the high pressure generated by the heart pumping blood into them.

Veins

Veins have a thinner wall and larger **lumen** (space inside them). The blood is under low pressure. To prevent blood flowing backwards, the veins have **valves**.

▲ Transverse section through an artery (left) and a vein (right) (× 26). The artery wall is thicker than the vein wall. It contains more muscle and elastic tissue. The lumen of the vein is bigger than that of the artery.

A How are arteries adapted for their function?

B How do veins prevent blood flowing backwards in them?

Capillaries

Capillaries are in tissues and organs. They form a network, so no cell is very far from a capillary. The blood flows slowly through capillaries and exchanges materials with the tissues. Some of the blood plasma is forced out through holes in capillary walls and it bathes the cells. It then passes back into the capillaries and then into the veins. Capillary walls are **permeable**.

- Glucose and oxygen pass (by diffusion) from this plasma into the cells, for respiration.
- Amino acids pass into the cells to be made into proteins the cell needs.
- Carbon dioxide and lactic acid pass from respiring cells into the plasma.

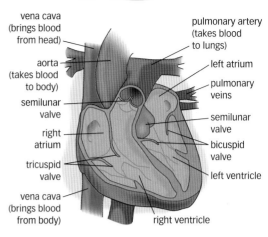

▲ This heart has been cut lengthways. The left side of the heart is on the right side of the page. Note the four main blood vessels associated with the heart: aorta, pulmonary artery, vena cava, and pulmonary veins. Blood pressure is greater in arteries than in veins.

The pump

Your heart is the pump of your circulatory system. Blood gets pumped through it twice. Blood goes from the left side of the heart to the body; back to the right side of the heart; from the right side of the heart to the lungs; back to the left side of the heart; and again to the body.

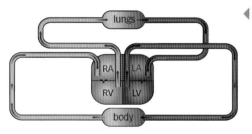

◀ Double circulation. The blood is pumped twice by the heart.

The left and right **atria** receive blood from veins. The left and right **ventricles** pump blood into arteries. The valves prevent backflow of blood.

The right side of the heart pumps blood to the lungs. The right ventricle has a thinner wall and generates less pressure so the delicate lungs are not damaged. The left side of the heart pumps blood to the rest of the body, including the head. Because the left ventricle wall is thick and muscular it can generate higher pressures, allowing fast delivery of oxygen to the body tissues and taking away waste quickly. At the same time as blood at higher pressures is pumped into the aorta to be pumped all over the body, blood at lower pressures is pumped into the pulmonary arteries to travel the shorter distance to the delicate lungs. This double circulatory system is an advantage for mammals.

Questions

1 Where is the tricuspid valve and what is its function?

2 Where are the semilunar valves? What is their function?

↓ C

3 Make a table to compare the functions of arteries, veins, and capillaries.

4 Why is the left ventricle wall of the heart thicker and more muscular than the right ventricle wall?

↓ A*

5 Make a table to compare the structure of arteries, veins, and capillaries.

6 Mammals have a double circulatory system. What does this mean?

Cells

All living things are made of one or more cells. Cells are very small and can only be seen with a light microscope. Plant and animal cells have some features in common, but plant cells have some features that animal cells do not have.

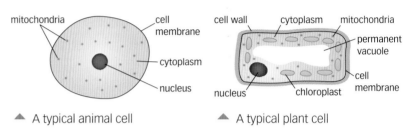

▲ A typical animal cell ▲ A typical plant cell

Plant and animal cells	Cell membrane	A thin layer around the cell. It controls the movement of substances into and out of the cell.
	Nucleus	A large structure inside the cell. It contains chromosomes made of DNA. The nucleus controls the activities of the cell, and how it develops.
	Cytoplasm	A jelly-like substance containing many chemicals. Most of the chemical reactions of the cell occur here.
	Mitochondria	Small rod-shaped structures that release energy during aerobic respiration.
Plant cells	Cell wall	A layer outside the cell membrane. It is made of cellulose, which is strong and supports the cell.
	Permanent vacuole	A fluid-filled cavity. The liquid inside is called cell sap. The sap helps support the cell.
	Chloroplasts	Small disc-shaped structures found in the cytoplasm of some plant cells. They contain the green pigment chlorophyll that traps light energy for photosynthesis. Chloroplasts are found in the cells of leaves and stems (where photosynthesis occurs) but not in the cells of roots or flowers.

Key words

stem cell, meristems

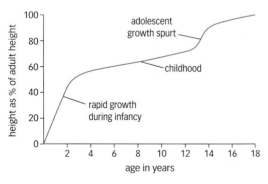

▲ A human growth curve

Bacteria consist of one cell. It is smaller than plant or animal cells and has no nucleus, no mitochondria, and no chloroplasts. Despite having no nucleus, bacterial cells contain DNA in the form of a single circular strand (chromosome) floating in the nucleus.

How do organisms grow?

Plants and animals start as one cell (a fertilised egg). It is an undifferentiated **stem cell**. It divides by mitosis to give cells which can then become specialised. The process of becoming specialised is called differentiation. Stem cells can develop into any of the different cell types and form tissues and organs.

Animals only grow in the early stages of their lives. All parts of them grow. They reach full size and stop growing.

Growth in humans

The graph on the previous page shows an average human growth curve. This is created by plotting the average height of a person as a percentage of the average full adult height, against their age. There are four main human growth stages:

- Infancy – growth is rapid following birth.
- Childhood – growth slows to a steady rate.
- Adolescence – puberty causes rapid growth.
- Adulthood – growth rate falls to zero.

Growth in plants

Plants grow and gain height throughout their lives by means of cell enlargement. Growth can be measured by increase in length, wet mass, or dry mass. However, plants only grow at specific parts, called **meristems**, at root tips, shoot tips, stem nodes, and buds. The meristems contain stem cells throughout their lives. These stem cells

- are undifferentiated
- have very thin walls
- are small and do not contain chloroplasts
- can divide, making new cells that differentiate.
- have very small vacuoles
- are packed very closely together

Differentiated plant cells cannot divide due to their thick rigid cell wall. They also have chloroplasts and a large vacuole.

Using stem cells

Scientists can get stem cells from spare very early embryos created during IVF treatment. Embryonic stem cells are still able to differentiate into any type of cell. Medical research is developing ways of using stem cells to

- treat Parkinson's disease
- grow tissues or organs
- repair spinal cord injuries
- treat type 1 diabetes.

Use of embryonic stem cells raises ethical issues as the spare embryos used could have developed into people. However, without stem cell research these embryos would still be discarded.

Embryonic stem cells are used in most stem cell research. Scientists can obtain adult stem cells from bone marrow or umbilical cord blood, which is less controversial than using embryonic stem cells. However, these stem cells cannot differentiate into as many different types of cells as embryonic stem cells.

Different growth rates

Human babies have undeveloped brains when born, otherwise their heads would be too large to pass along the birth canal. Just after birth the child's brain is growing faster than the whole body. During adolescence your reproductive organs grow a lot as you become an adult.

Measuring growth

Measuring increase in dry mass is best, because wet mass varies according to how much water is present in the tissues. However, organisms must be killed to calculate their dry mass. Using length as a measure of growth is simple, but won't account for the variable growth rates of different parts of an organism.

Questions

1. What is cell differentiation?
2. State the two phases of rapid growth (growth spurts) in humans.
3. What are meristems?
4. Make a table to compare growth in plants with growth in animals.
5. Discuss the ethical issues raised by using human stem cells in medical research.
6. Explain the difference between adult and embryonic cells.

▲ Oilseed rape (*Brassica napus*). This crop has been genetically modified by selective breeding to prevent it producing an oil that is toxic to humans, as we obtain edible oil from this plant. However, if we want to grow these plants for biofuel, the toxic oil would be better as it would release more energy.

A How did the process of selective breeding start?

B What is cross breeding?

Did you know...?

Potatoes were introduced into Europe from South America during the sixteenth century. Many people viewed them with suspicion at first. However, they are nutritionally superior to wheat. One acre of potatoes produces four times as much energy content as one acre of wheat, and potatoes contain more nutrients. There are many different varieties.

Selective breeding of plants

Humans began to practise agriculture 10 000 years ago. They saved seeds from plants that had desired characteristics, to grow the following year. In this way wild grasses became genetically modified and evolved into the cereal plants such as wheat and barley that we know today.

Improvements in staple crops, such as maize, wheat, millet, rice, and potatoes, are important. These crops form the bulk of people's diets. We need varieties of crops that

- give a high **yield** (produce a lot of the edible plant part)
- are resistant to diseases that may cause health problems within the species
- do not bend over and break their stalks in the wind
- depending on where they are grown, are resistant to drought/flooding/frost
- taste good
- have a long shelf life
- contain desired amounts of particular nutrients.

A **selective breeding** programme to improve crop plants can take up to 20 years.

- Parent plants with the desired characteristics are selected. One may have a high yield but be susceptible to a disease. The other may have a lower yield but be resistant to the disease.
- These are **cross bred** – pollen from one parent plant is placed on the female parts of the other parent plant.
- The seeds are collected and grown.
- The offspring that have inherited both characteristics – higher yield and some resistance to the disease – are selected and cross bred again.
- Their seeds are collected and grown.

The process is repeated over many generations, until a new variety is produced with all the desired characteristics.

Selective breeding of animals

Humans domesticated wolves, which eventually evolved into dogs, between 15 000 and 30 000 years ago. The wolves got food; humans got protection, companionship, and help with hunting.

When humans started farming they also began to domesticate animals, such as

- sheep and goats about 11 000 years ago
- pigs about 9000 years ago
- cattle about 8000 years ago
- horses 5000 years ago.

Humans used the animals for meat, milk, bones, wool, hides, dung, and to do work. They chose the most docile and manageable animals.

In more recent times, humans have started carrying out selective breeding programmes by selecting parents with desired characteristics, breeding from them, selecting the best offspring and breeding from them, over many generations. This produces breeds of animals that

- have more muscle and less fat for lean meat
- produce higher milk yields
- lay more eggs
- reach maturity quicker
- have better/more wool
- can run faster (such as racehorses and greyhound dogs).

Disadvantages of selective breeding
In the selective breeding process, inbreeding may reduce the gene pool. This could lead to

- an accumulation of harmful recessive characteristics leading to health problems
- reduction in variation.

▲ These Zebu cattle can tolerate hot, dry climates. They are used in African countries for their meat, milk, blood, leather, and to do work.

Key words

yield, selective breeding, cross breeding

Exam tip

✔ Remember that new evidence is always being found for facts, such as how long ago dogs were domesticated. Sources of information may differ slightly. Whatever information you are given in an exam question, use that data to answer the questions.

Questions

1 Humans carry out selective breeding programmes with animals. State four characteristics that have been selected in animals.

2 Describe a selective breeding programme to produce a tomato plant with edible fruits that will grow well in the UK, from two parents: one with poisonous fruits that grows in cool climates, and one with edible fruits that grows in tropical regions.

↓ E

3 Describe a selective breeding programme to produce a breed of chicken that lays one large egg almost every day of the year.

4 Explain the possible disadvantages of selective breeding in animals and plants.

↓ A*

Learning objectives

After studying this topic, you should be able to:

✔ recall that genetic engineering can artificially transfer genes from one organism to another, to produce organisms with desired characteristics

✔ understand the difference between gene therapy and genetic engineering

A What is genetic engineering?

B What are the advantages of using GM bacteria to make insulin?

What is genetic engineering?

Genetic engineering is a faster way of genetically modifying organisms without going through a selective breeding programme. Organisms produced by genetic engineering are described as **genetically modified**.

Enzymes are used to cut DNA, to obtain a gene for a desired characteristic, and to insert that gene into another organism's DNA. This can produce organisms with different characteristics.

Some examples of genetic engineering
Making human insulin

Insulin to treat people with diabetes used to be obtained from pig pancreases. Now, the human gene for making insulin is inserted into a bacterium. The bacteria make insulin (a protein). Scientists can collect it.

Advantages:
- Enough insulin can be made to treat all the people with diabetes.
- There is no risk of transferring diseases from pigs to humans.
- People with diabetes who are vegetarian will not object.

Genetically modified (GM) crops

GM crop plant	Use
Golden rice	Genes that control the production of beta-carotene have been taken from daffodils and put into rice. The GM rice plants make beta-carotene in the rice grains. When humans eat this rice they turn beta-carotene into vitamin A. Rice is the major part of the diet in many developing countries. Non-GM rice grains do not contain beta-carotene. Each year in the world 600 000 children go blind or die due to lack of vitamin A, as there are not many green vegetables available for them to eat. Golden rice will be a good way of providing enough vitamin A to children in developing countries at no extra cost.
Cotton	Cotton fibres are used for textiles, and the seeds provide oil and protein for animal feed or oil for margarine. GM cotton is resistant to caterpillar pests. Cotton plants used to be sprayed with chemical pesticides but often these did not kill the caterpillars, which were inside the seed capsule (boll). GM cotton has a gene from a bacterium. The gene codes for a toxin that kills the caterpillars. This Bt toxin has been used for decades by extracting it from bacteria. Now the GM cotton plants themselves make the toxin and kill the caterpillars even when they are inside the boll.

Many other GM crops are grown in Canada, the USA, India, China, South America, Kenya, Mexico, and Australia.

- 77% of all soya is GM and resistant to herbicide (weedkiller), so spraying the crop with weedkiller kills only the weeds.
- 80% of maize grown in the USA is GM for resistance to an insect pest.
- GM bananas, a staple crop in Kenya, are resistant to disease and also contain more nutrients.
- GM tomatoes have had a gene from a cold-water fish inserted into them to make them frost-resistant.

Gene therapy

Changing a person's genes in an attempt to cure disorders is called **gene therapy**.

Gene therapy does not change an organism's genes permanently. Copies of a functioning (normal) gene or allele may be inserted into certain body cells of a person who has a recessive genetic disease, such as cystic fibrosis.

If the genes were inserted into a gamete or a zygote, then all the cells of the new individual would have the healthy genes. This would be genetic engineering. This is not done with humans. There are ethical guidelines and laws to prevent it.

Questions

1 How does gene therapy differ from genetic engineering?

2 What do you think are the advantages of producing new varieties of crop by genetic modification rather than by selective breeding?

3 Scientists are developing a type of GM corn that contains fish oils. These oils protect us from heart disease and help brain development. Fish get them from the algae they eat. Genes from the algae can be put into corn. Why do you think this may be particularly useful in the future?

↓ E
↓ C
↓ A*

Did you know...?

Nature has its own genetic engineer. The bacterium *Agrobacterium tumefaciens* has been inserting some of its genes into plants for a very long time. Scientists use it as a tool. They insert genes into it and let the bacterium carry these genes into certain plants.

Key words

genetic engineering, genetically modified, gene therapy

Principles of genetic engineering

The main principles of genetic engineering can be summarised as:

- selection of desired characteristics
- isolation of genes responsible
- insertion of the genes into other organisms
- replication of these organisms.

Exam tip OCR

✔ Remember that gene therapy does not cure but treats some genetic diseases. It does not treat dominant disorders because we cannot cut out harmful genes and replace them. Try not to talk about 'replacing genes'.

🔺 A researcher at a research station in the UK compares the growth of GM crops with non-GM crops. These trials find out if GM crops will harm the environment.

A Why are people worried about growing GM crops?

B How is it useful to scientists that GM crops are allowed in the USA but not Europe?

Why are people concerned?

GM technology can rapidly produce organisms with desired features. However, Many people have worries that inserting genes into GM crops will have unexpected side-effects. However, in the USA people have been eating GM crops since the early 1990s. This has produced a good natural 'experiment'. The control group is in Europe, where GM crops are not grown commercially and GM food is not sold.

In the USA:

- No superweeds have developed that are resistant to weedkillers.
- There is no reported reduction in biodiversity where GM crops are grown, compared with where they are not grown. Fears that the Monarch butterfly would become extinct have not proved true so far.
- No one has reported a health problem from eating GM food.

Many people are still developing health problems from eating non-GM food, such as processed foods and foods containing a lot of saturated fat and salt, or from eating too much.

Weighing the risks

Novel foods such as GM products are tested. Golden rice has to be tested to see if it will cause allergies, now that it has beta-carotene in it.

For GM	Against GM
More people can be fed, as GM crops produce higher yields.	Poor farmers that would benefit from the high-yield GM crops may not be able to afford the seeds.
Many people have no problem with the idea of eating GM produce.	Those who have a problem with the idea of eating GM produce won't buy it, meaning that farmers lose money and markets.
GM farms may have increased productivity while using fewer inputs, so food costs may fall.	GM crops may change the ecosystem in ways that cannot be reversed.
Most GM crops have been safe so far.	GM crops could cross-pollinate with wild plants, which could lead to unexpected side effects.

In 1991, Sainsbury's sold tomato puree made from GM tomatoes. It was thicker than non-GM puree and it sold well. However, because some people were concerned, supermarkets stopped selling GM products. Many tabloid newspapers have highlighted these fears and have given high profile coverage to concerns over the benefits of GM. Arguments for and against GM products are compared on the previous page.

In the developing world one billion people are starving and another one billion are on the brink of starvation. In the developed world we do not currently have a food shortage, and we do not go blind due to lack of beta-carotene, so many people do not see the advantages of GM crops. They therefore think the potential risks outweigh the advantages. A balanced view needs to consider people in all parts of the world.

The potential benefits of GM

GM crops can be part of the solution to feeding the growing world population, which will probably increase by 50% to 9.3 billion by 2030. India and China are becoming richer, and their people want to eat more meat. So grain production will need to double to feed both the people and the extra livestock.

During the 1960s and 1970s there was a green revolution. Selective breeding has modified crop plants to give greater crop yields. In addition, more **fertilisers** and **pesticides** were used. This all helped boost world food production. But there was a price to pay:

- Many farmers applying the pesticides have become ill or died as a result of exposure to them.
- Using a lot of fertiliser has damaged the soil.

In addition to this, global warming and water shortages in many areas limit the growth of crops.

Crops that are resistant to pests could be developed, allowing us to use fewer pesticides. This would be less harmful to useful insects in the environment. Higher yielding crops may reduce the need for fertilisers. Drought resistant crops could reduce the need for water. GM crops will not be the only solution, but they could play an important part in the future of agriculture.

▲ Tomato puree made from GM tomatoes

Key words

fertiliser, pesticide

Questions

1 Explain why, if the world population increases by 50%, we will need to grow twice as many grain plants.

2 State four potential benefits of GM crops.

3 State three potential risks associated with GM crops.

4 Everything has a risk attached to it. We always weigh up the risks to see if the benefits are great enough to justify them. We all do this every time we get into cars, take a plane flight, or cross the road. With few problems reported in the USA, explain why some people in the developed world say they are concerned about the risks of GM crops.

Learning objectives

After studying this topic, you should be able to:

✓ recall that cloning is an example of asexual reproduction and produces genetically identical organisms

✓ know that plants can be cloned by cuttings or by tissue culture

✓ recall that animals have been cloned by nuclear transfer

Asexual reproduction in plants	
Strawberry plant	Produces special stems called runners. New plants, clones of the parent, develop at the end of each runner.
Spider plant (*Chlorophytum*)	Produces runners like the strawberry plant. The new plantlets can be planted.
Potato plant	Produces tubers (swollen parts of underground stems). If not eaten, each potato tuber could produce a genetically identical new plant.

▲ These identical twin brothers are naturally occurring clones of each other. They are genetically identical because they are from one fertilised egg that split into two.

Asexual reproduction

Some organisms may reproduce asexually, using mitosis.

- Only one parent is needed.
- There are no gametes.
- There is no mixing of genetic information.
- The offspring are genetically identical to each other and to the parent – they are **clones**. The offspring have the desirable characteristics of the parent.
- Growers take cuttings from plants. They cut off a bit of stem or root, and grow it into a new plant which is a clone of the parent. Tissue culture can also be used to clone plants.
- As the new plants have no genetic variation, they would all be susceptible to environmental changes or particular diseases.

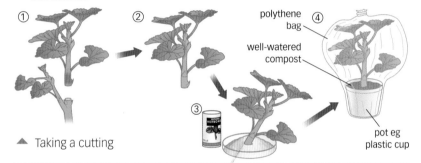

▲ Taking a cutting

Tissue culture

Tissue culture can be used to make new plants by asexual reproduction. Technicians take many small pieces of tissue from a plant with desirable characteristics and put them into special sterile liquid or jelly. This very clean **aseptic technique** ensures that bacteria or moulds do not contaminate the cultures. The culture medium contains some nutrients and special chemicals. The plant cells are kept at a suitable temperature and light, and develop into new plants with roots, leaves, and shoots.

Many plants that are difficult to grow from seed can be grown using tissue culture or cuttings.

Cloning plants is easier than cloning animals, because mature plants still have stem cells in their meristems. Mature animal cells can no longer differentiate. However, Dolly showed that all the genes in an adult animal cell nucleus can be switched on again.

Dolly the sheep

Dolly was the first mammal cloned from an adult cell.

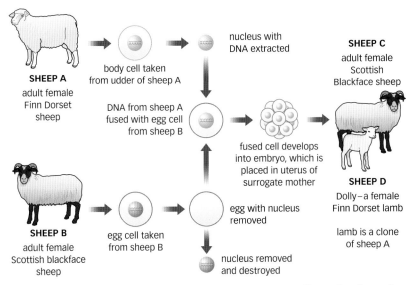

- An unfertilised egg was taken from a ewe (female sheep) and its nucleus was destroyed.
- A cell was taken from the udder of a different ewe.
- Its nucleus was implanted into the empty egg.

> - An electric shock was given to the resulting egg cell to make it divide.
> - It developed into an embryo that was put into a surrogate mother sheep.

The resulting cloned sheep (Dolly) was a **nuclear transfer clone** of the sheep from which the udder cell came.

Scientists have made genetically engineered sheep that make useful medicines for humans in their milk. These sheep can breed, but their offspring may not inherit the human gene. Also, half of their offspring would be male and not make milk. If these sheep could be mass produced, then many sheep able to make the medicine could be created.

Human cloning

Spare eight-cell embryos can be split and allowed to develop into cloned embryos. Cells from these can be used for stem cell research. They do not develop into babies.

> In the future, organs for transplants may be grown from stem cells.

Future transplants

Soon, human trials using tissue from specially bred modified pigs could begin to treat patients with diabetes, Huntington's disease, Parkinson's disease and blindness. Replacement organs from the pigs may be transplanted into humans. Genetically modifying the pigs may overcome rejection problems. There are fears that viruses may pass from pigs to humans so experts say trials should be closely monitored.

Ethics of cloning

Potential benefits:
- Creation of replacement organs and tissues.
- Could allow infertile parents to have children.
- Extending life by replacing ageing tissues and organs.

Potential issues:
- Humans created as tools or products for medicine.
- Clones would be identical twins of the cell donor.
- Research to perfect cloning could lead to damaged clones.
- Decreasing genetic diversity caused by asexual cloning.

Questions

1 A plant grower has a variety of geranium that sells well. Should he use cuttings or seeds to grow lots of them? Explain your answer.

2 Sometimes new cuttings are placed in pots covered with clear plastic bags. Why?

3 What is adult cell cloning?

4 Discuss the social and ethical issues of animal cloning.

Module summary

Revision checklist

- Chromosomes are made of DNA and carry an organism's genetic code. Each gene is a length of DNA that has specific coded instructions to make a protein.

- Skin, hair, muscle and bones are made of protein. Enzymes, antibodies and haemoglobin are also proteins.

- Enzymes catalyse chemical reactions in cells. Each type of enzyme only works with a specific substrate and works best at a particular temperature and pH.

- Mutations in DNA may lead to different proteins being made. Mutations can be harmful, neutral, or beneficial.

- Respiration in mitrochondria in cells provides the energy needed for an organism's life processes.

- Respiration rate increases during exercise, so pulse and breathing rate increase.

- Body cells and some single-celled organisms divide by mitosis. Mitosis produces two daughter cells that are genetically identical to each other and to the parent cell.

- Cells divide by meiosis to make gametes for sexual reproduction. Meiosis produces genetic variation.

- Blood is a liquid tissue. It contains red cells that carry oxygen and white cells for defence. The watery plasma carries digested food, waste and hormones.

- Blood leaves the heart in arteries and returns to it in veins. At capillaries materials are exchanged with cells.

- Organisms grow through cell division. They start as one unspecialised stem cell which then differentiates.

- Selective breeding can produce organisms with desired characteristics, as can genetic engineering.

- Genetic engineering involves transferring genes from one species of organism into another. Gene therapy involves inserting genes from one individual into another of the same species. Gene therapy may treat some genetic disorders.

- Genetically modified crops have many potential benefits, but some people fear they may have unexpected side-effects.

- Cloning is a type of asexual reproduction. Cloning occurs in nature (strawberry runners, identical twins); animals and plants can be cloned in laboratories. Embryo cloning could produce stem cells for medical treatments.

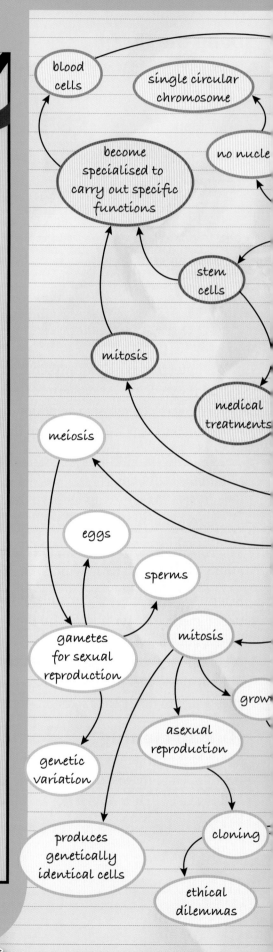

NOW USE THE B3 GRADE CHECKER ON PAGE 244

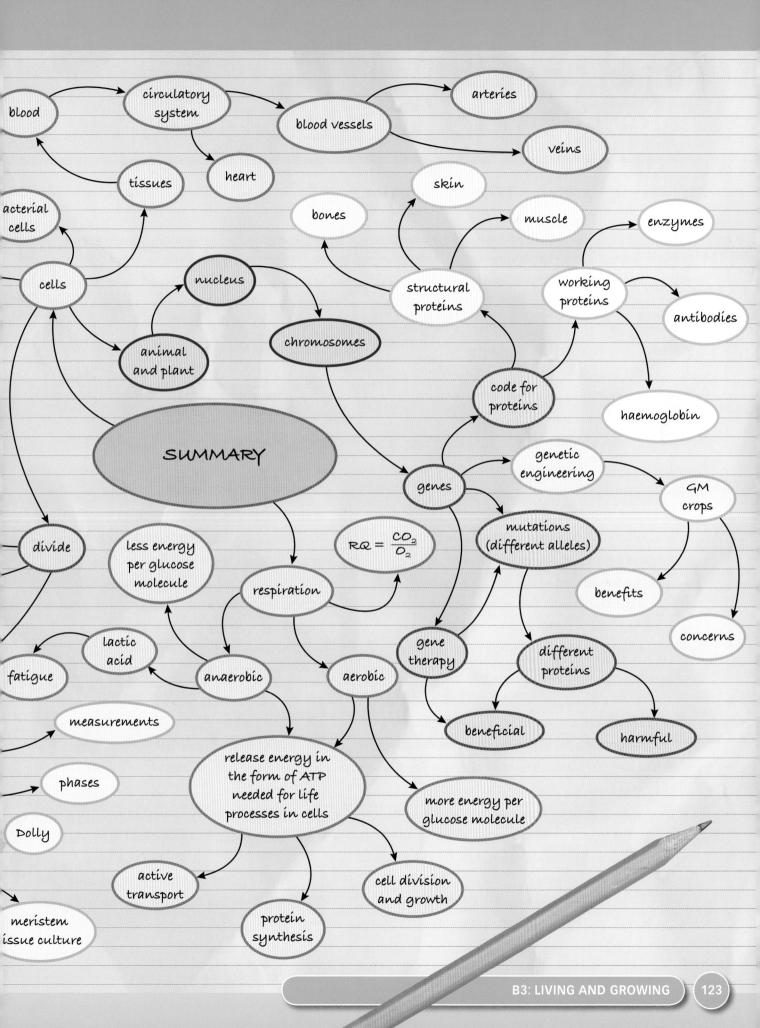

blood

circulatory system

arteries

blood vessels

veins

heart

tissues

skin

bones

muscle

enzymes

acterial cells

cells

nucleus

structural proteins

working proteins

antibodies

animal and plant

chromosomes

code for proteins

haemoglobin

SUMMARY

genes

genetic engineering

GM crops

divide

less energy per glucose molecule

$$RQ = \frac{CO_2}{O_2}$$

mutations (different alleles)

benefits

concerns

respiration

lactic acid

fatigue

anaerobic

aerobic

gene therapy

different proteins

measurements

beneficial

harmful

phases

Dolly

release energy in the form of ATP needed for life processes in cells

more energy per glucose molecule

meristem issue culture

active transport

protein synthesis

cell division and growth

OCR gateway *Upgrade*

Answering Extended Writing questions

QUESTION

What features of animals and plants might be selected for in a genetic engineering programme? Discuss the benefits and risks of genetic engineering. Explain how gene therapy differs from genetic engineering.

The quality of written communication will be assessed in your answer to this question.

Bacteria can be geneticly ingineerd so they make insulin for people with diabetes. People may have designer babies. scientists shoudnt play god. gene therapy is when a faulty gene is replaced with a good gene.

Examiner: There is one example of a genetically engineered organism here, but it is about bacteria instead of animals and plants. The second sentence is not relevant. Gene therapy does not replace genes. Poor spelling, punctuation, and grammar.

Some sheep are genetically engineered to make useful things in their milk. Some GM plants grow more and have more yield. It may be dangerous to eat food made from GM plants. GM plants may cross with other plants. Gene therapy is when you put a healthy gene into cells. This can treat genetic diseases. It has to be repeated.

Examiner: The advantages and possible disadvantage of genetic engineering are vague. Gene therapy is quite well described, but the difference between it and genetic engineering is not clear. There is no explanation of what genetic engineering is. The spelling, punctuation, and grammar are good.

Genetic engineering is changing an organism's genes. Genes are put in from another organism. Usually a different species. This changes the features of the organism. For example plants that resist pests or frost or plants with better nutrients, like Golden rice. This could stop lots of people going blind in India and Africa. Some people think genetic engineered plants could lead to superweeds.

Gene therapy does not permanently alter people's genes. It can treat diseases like cystic fibrosis, but the people can still pass the faulty gene to their children.

Examiner: A well organised answer. It clearly explains what genetic engineering is and gives examples of three features that may be selected for. However, no animal examples are given. A possible advantage and disadvantage of GM plants is included. The candidate shows how gene therapy is different from genetic engineering and that gene therapy can be used to treat some genetic diseases.

Exam-style questions

1 The diagram shows some blood as seen under a light microscope.

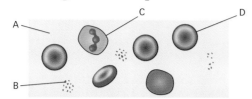

A01 a Name parts A–D.

A01 b Match each part with the correct function: from the list below:

carries oxygen; defence; blood clotting; carries hormones and waste

A01 c Blood is an unusual tissue because
  i it contains various cell types
  ii it is found only in animals
  iii it is liquid
  iv it is made in the body.

A01 d Fill in the gaps. The first letters have been given.

Blood leaves the heart in vessels called a_____ and returns to the heart in vessels called v_____. At the body tissues substances are exchanged between small vessels called c_____ and the body cells.

2 a Why do cells need oxygen?

A01 b Which parts of cells use oxygen?

A01 c Write a word equation for anaerobic respiration.

A01 d Which type of respiration releases more energy from each glucose molecule?

A01 e Which type of respiration is used by an athlete running a 100 m sprint?

A01 f Explain why an athlete's heart rate and breathing rate are higher for a few minutes after a sprint.

A01 g When do plants respire?

A02 h Yeast is used in bread making. It respires and produces carbon dioxide to make bread rise. Explain why more carbon dioxide is produced by the yeast at room temperature than when cold.

3 a Explain how a gene codes for a protein.

A01

A02 b Scientists have genetically modified rice by placing a gene that controls the production of beta-carotene (vitamin A) into it. Many people in Asia rely on rice as their main source of food. Explain why this genetically modified rice could help people in Asia.

A01 c Some people are worried about genetic engineering. Describe one possible reason for their concern.

Extended Writing

4 Explain why living organisms need
A01 energy.

5 a Explain how mutation can lead to
A01 different proteins being made.

b Anil says that all mutations are harmful. Jamil disagrees. Explain why Jamil is correct.

6 Evaluate the pros and cons of using
A02 genetically modified pigs as a source of tissues and organs to treat human patients.

A01   Recall the science

A02   Apply your knowledge

A03   Evaluate and analyse the evidence

B4
It's a green world

Why study this module?

Photosynthesis is one of the most important biological processes. It is through photosynthesis that energy is trapped into the living world. Once trapped, this energy is used to power the entire living world in all its glory.

In this module, you will study photosynthesis as a process, and look at where the process occurs. You will also look at processes for sampling the distribution of living organisms in the environment, and identifying them.

You will examine the processes by which substances are transported around plants, and the importance of soil minerals in the healthy growth of plants. The return of these minerals to the soil through decay will also be considered.

This module also reviews the methods of food preservation that are economically important in the food industry, together with various approaches to food production in farming.

You should remember

1 The environment can be studied, by sampling the distribution of organisms.

2 Plants make food by the process of photosynthesis.

3 Photosynthesis occurs in the leaves.

4 Diffusion is the movement of particles from a high concentration to low concentration.

5 The carbon and nitrogen cycles, and recycling.

The space age comes to our farms… Modern farming technology has brought some amazing new approaches to the age-old process of growing crops. Attempts to grow more crops per unit of land have led to the development of hydroponics. The photograph shows lettuces being grown in hydroponic conditions – soil has been abandoned, and the plants are suspended with their roots in a nutrient-rich solution. This means that many plants can be grown quickly, as the ideal growing conditions are supplied. These techniques allow farmers to grow crops on land which might not otherwise be suitable for farming. Also, the plants are grown indoors, so can grow at any time of year. This method of farming could be used in space, to feed colonies on the moon and beyond.

▲ Students count how many organisms of a certain species are inside the quadrat. This gives a sample.

▲ Collecting insects with a pooter

There's a lot out there!

When biologists investigate where organisms live, they meet problems:

- There are very many different organisms.
- They seem to live all over the place.

It is difficult to make sense of the huge amounts of data.

To overcome these problems, biologists have devised a series of techniques to collect information about two things. First, they record the location of organisms of one species; this describes their **distribution**. Second, they record the number of organisms of a particular species in an area; this is the **population**.

Different populations live together in one area, and together they form a **community**. Biologists look for **relationships** between the organisms in a community by studying how their distributions overlap. They also study how factors in the environment affect their distributions.

To collect this information biologists need techniques to

- collect organisms
- count the number of organisms in each species
- record where the organisms are found
- collect accurate data
- collect the data fairly
- collect reliable data.

Biologists use a technique called **sampling**. This means counting a small number of the total population and working out the total from the sample.

Sampling techniques

1. Quadrats are square frames of a standard area. They are put on the ground to define an area. The numbers of organisms of particular species in the frame can then be counted.
2. Transect lines are tapes that are laid across an environment. You can count the organisms that touch the tape, such as plants on the ground, to study their distribution. Alternatively you can lay quadrats at regular intervals down the tape to record the distribution of the organisms inside.
3. Nets are used to catch animals such as butterflies or fish, allowing you to count and record numbers of animals.

4. Pooters are containers with a straw device, used to suck in small animals so that they can be identified and counted.

5. Pitfall traps are small containers buried in the ground which collect small animals, allowing you to sample the animals in the area.

When you have enough readings, it is possible to make estimates of the size of a population from your sample. This is done by scaling up from a small sample area. The technique of 'capture–recapture' is used.

1. Capture a sample of organisms in an area (first sample), count, mark, and release them.

2. Recapture and count a second sample of organisms in the same area at a later date.

3. Count the total number of recaptured (previously marked) organisms in the second sample.

Population size can then be estimated using this equation:

$$\frac{\text{number in first sample} \times \text{number in second sample}}{\text{number in second sample previously marked}}$$

Sampling accurately

A big enough sample

The apparatus should allow you to count a reasonably large number of the type of organism you are studying. For example, if you use too small a quadrat, you will record fewer plants and animals. A small sample size is not very accurate, as small samples are affected more by stray results.

Being reliable

Repeat readings make the data more reliable. If only one quadrat is recorded, it might not represent the population accurately. The more quadrats you record, the more reliable your data will be.

Being fair

To be fair, all your readings should use the same equipment. They also need to be placed fairly. When recording distribution, quadrats can be placed at regular intervals along a transect. This avoids you choosing places that look promising, which would give biased readings. When estimating population size, quadrats should be placed randomly in an area, rather than choosing where to place them.

> **A** How could a group of students record the distribution of limpets down a beach?

When estimating the size of a population using capture–recapture data, biologists assume that no death, immigration, or emigration has occurred within the population. They also ensure that:

- the sampling method used is the same each time
- the marking of organisms does not affect their survival.

This ensures that population estimates are as accurate as possible.

Exam tip

✔ Useful memory aids for sampling are 'Accuracy using Apparatus' and 'Reliability needs Repeats'.

Questions

1 Name three devices used to collect small animals. ↓ E

2 Describe what techniques you would use to estimate the population of daisies in a school field. ↓ C

3 Explain why you think that collecting data using sampling techniques gives only a rough estimate of population size. ↓ A*

Learning objectives

After studying this topic, you should be able to:

✔ know that there is a great variety of organisms

✔ use keys as a tool to identify organisms

Key words

habitat, key, ecosystem, biodiversity, zonation

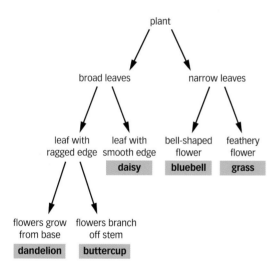

▲ A spider key for identifying common plants

A What is a community?
B What is a habitat?
C What is a key?

Who's who?

All the different plants and animals in an area make up the community. But there are very many plants and animals in the world. It can be difficult to identify them, and describe them accurately to other biologists. This is true when we study any **habitat** where lots of plants and animals live.

Biologists have developed systems called **keys** to help them identify plants and animals.

Keys

Keys are based on a series of questions that look at visible features or characteristics of the organisms. There are two common types of key:

- spider keys
- numbered keys.

Spider keys

In a spider key, answers to questions take the reader along one of two branches. For example, to identify a plant you first answer a question. Your reply takes you along one branch of the key. Then you answer a second question, which again takes you along one of two branches. The questions will continue until you have named the plant. This type of key looks like a spider diagram, giving it its name.

Numbered keys

Spider keys are easy to follow, but they get very messy when there are more than five or six organisms. Numbered keys ask the same type of questions, but in a list. As you answer each question, you are sent on to another numbered question.

1	Leaves narrow	go to 2
	Leaves broad	go to 3
2	Bell-like flowers	Bluebell
	Feathery flowers	Grass
3	Leaves with ragged edge	go to 4
	Leaves with smooth edge	Daisy
4	Flowers grow from base of plant	Dandelion
	Flowers branch off stem	Buttercup

Natural ecosystems

An **ecosystem** includes all the living things in an area (the community), and how they interact with each other and the physical conditions around them. An ecosystem is self sufficient, which means that it needs nothing supplied to it except energy from the Sun. Natural ecosystems, such as woodlands, lakes, and seashores, often contain many different types of organism. We say they have large **biodiversity** which is a variety of different species in a habitat.

Case study: the seashore

Organisms on a beach are not randomly placed. Why is this? There are two reasons:

- The effect of physical factors.
- The effect of other organisms.

Biologists wanted to examine the distribution of two species, mussel and barnacles, on a beach. They set up a transect line on the beach and look regular counts of the animals using quadrats. They displayed their data in the kite diagrams on the right.

> The diagrams show that a zone of barnacles gradually gives way to a zone of mussels as you move down the beach. The two species live in different bands, or zones, of the beach. This is called **zonation**. The zonation here is due to the gradual change in the amount of time the different areas of the beach (and therefore the organisms that live there) are exposed to the air when the tide is out. This a physical factor in the distribution of these organisms.

Competition between the two species is another reason for the distribution of barnacles and mussels shown in the diagrams. The barnacles cannot grow so well further down the beach, not because conditions are too wet but because they cannot compete with the mussels for food.

Artificial ecosystems

Artificial ecosystems are those created by humans, such as fish farms, forestry plantations, and gardens. They usually have fewer types of organisms than natural ecosystems, and therefore have a lower biodiversity.

> Humans often control what organisms live in an artificial ecosystem, and remove all unwanted species.

▲ Barnacles

▲ Kite diagram for distribution of barnacles

▲ Mussels

▲ Kite diagram for distribution of mussels

Questions

1. Explain why numbered keys can be more useful to biologists than spider keys.
2. What are the questions in a key based on?
3. Explain why biologists need to be able to identify organisms.
4. Explain why there are fewer grass plants as you walk into a woodland.

Learning objectives

After studying this topic, you should be able to:

✔ know that photosynthesis is the process by which plants make their own food

✔ appreciate the source of the raw materials for photosynthesis

✔ understand the fate of the products of photosynthesis

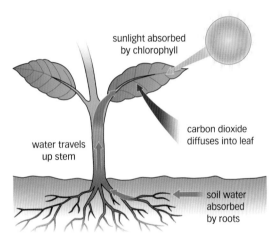

▲ In photosynthesis the plant uses sunlight energy to convert water and carbon dioxide into carbohydrates

---

A What are the two raw materials a plant needs for photosynthesis?

B What else does a plant need in order for it to photosynthesise?

C Explain why humans could not survive without photosynthesis.

Feeding in plants

Plants do not take in ready-made food like animals do. They have to make their own food. To do this, plants take in
- carbon dioxide from the air through pores called stomata
- water from the soil through root hairs.

Plants trap the Sun's energy in a substance called **chlorophyll**, which is in the chloroplasts in their cells. They use this energy to build up the carbon dioxide and water into glucose and oxygen. This process is called **photosynthesis**.

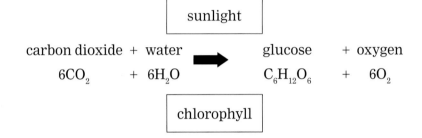

Photosynthesis occurs in two stages. The first uses light energy to split water into waste oxygen gas and hydrogen ions. The second stage combines the hydrogen ions with carbon dioxide to form glucose.

What does the plant make in photosynthesis?

You can see from the equations that there are two products of photosynthesis:

1. Food: this is glucose, a carbohydrate. Some is used for respiration in the plant's cells. The rest is converted to other substances and stored in the plant.

2. Oxygen: this is a waste gas produced in photosynthesis. Some is used for respiration in the plant's cells. The rest is given off through the stomata into the plant's surroundings. Without plants there would be no oxygen in the air for animals to breathe.

Converting glucose to other substances

The glucose produced in photosynthesis can be converted to other substances that the plant needs. For example, it may be used to make the sugar sucrose, found in sugar cane.

If it is not used, the glucose can be changed into insoluble starch and stored. Stored starch can be used for respiration at night, when there is no sunlight and the plant is not making glucose by photosynthesis. The glucose made in photosynthesis is converted to sucrose to be transported around the plant to parts that need it. Sucrose is good for transport because it dissolves in water and flows easily.

Plants are not made of sugars alone. The plant converts sugars to other substances such as cellulose, proteins, and fats, which it needs to grow and for other functions.

Storing glucose

Glucose is stored in the plant as starch. This has three advantages:

1. Starch can be converted back into glucose for respiration in plant cells.

2. Starch is insoluble and so will not dissolve in water and flow out of the cells where it is stored.
3. Starch does not affect the water concentration inside the cells.

Photosynthesis and respiration

Plants do not only photosynthesise. They also respire all the time because respiration releases energy needed by the plant to grow and survive. This affects the movement of gases into and out of the leaf.

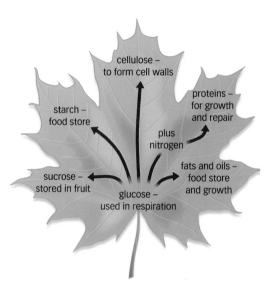

▲ Glucose from photosynthesis is converted to all the substances that a plant needs

Key words

chlorophyll, photosynthesis

Daytime	Night-time
Plants photosynthesise at a faster rate than they respire, taking in more carbon dioxide for photosynthesis and releasing the extra oxygen produced.	Plants respire but do not photosynthesise. Oxygen is taken into the leaf and waste carbon dioxide is released.

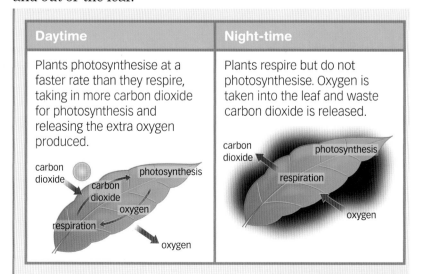

Questions

1. Where does the energy for photosynthesis come from?

2. Plants convert some glucose to cellulose. What is the cellulose used for?

3. Explain why plant cells store carbohydrate as starch.

▲ The rate of photosynthesis in this glasshouse is increased using artificial lighting

Key words

rate of photosynthesis,
limiting factor

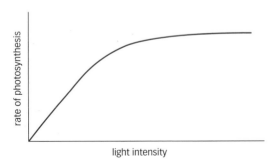

▲ Graph to show how the rate of photosynthesis changes as light intensity increases

The growing season

Plants do not grow at the same rate all year round. Most plants grow best in the spring and summer. This is when the conditions for growth are best. In spring and summer, the weather is usually warmer and there is more sunlight. These conditions are good for photosynthesis and therefore for growth, because the light energy is needed for photosynthesis, and the warmth speeds up the reactions of photosynthesis.

Increasing the rate of photosynthesis

The **rate of photosynthesis**, or how quickly the plant is photosynthesising, depends on several things. The following factors will speed up photosynthesis:

- more carbon dioxide
- more light
- a warm temperature.

People who grow plants commercially in a glasshouse try to make sure their plants have the best conditions. They use lighting systems which increase the hours of daylight available to plants, and they use heaters that burn gas or other fuels to add warmth and release carbon dioxide.

> **A** List three things that will increase the rate of photosynthesis.
>
> **B** Why do you think British woodland flowering plants such as bluebells flower in May?

Factors affecting the rate of photosynthesis

The rate of photosynthesis may be limited by the following factors.

Availability of light

Light provides the energy to drive photosynthesis. The more light there is, the faster the rate of photosynthesis. This is true provided that there is plenty of carbon dioxide, and the temperature is warm enough.

Amount of carbon dioxide

Carbon dioxide is one of the raw materials for photosynthesis. The more carbon dioxide there is available, the faster the rate of photosynthesis. (Again, this is only true if there is plenty of light and a suitable temperature.) Carbon dioxide is often the factor in shortest supply, so it is often the limiting factor for photosynthesis.

A suitable temperature

Temperature affects how quickly enzymes work. Enzymes make the reactions of photosynthesis happen. As the temperature rises, the rate of photosynthesis increases (providing there is plenty of carbon dioxide and light). However, if it becomes too hot the enzymes will be denatured and photosynthesis stops.

Limiting factors

When a process is affected by several factors, the one that is at the lowest level will be the factor which limits the rate of reaction. This factor is called the **limiting factor**.

If the limiting factor is increased, the rate of photosynthesis will increase until one of the other factors becomes limiting. For example, if photosynthesis is slow because there is not much light, giving the plant more light will increase the rate of photosynthesis, up to a point. After that point giving more light will not have any effect on photosynthesis, because light is no longer the limiting factor. The rate may now be limited by the level of carbon dioxide, for example.

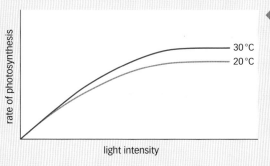

◀ Graph to show the effect of increasing light at a temperature of 20°C

Light levels are limiting initially. The rate of photosynthesis then levels off. It increases at a higher temperature, so at higher light levels, temperature becomes the limiting factor.

▲ Warm, sunny conditions mean light and temperature are not limiting factors for photosynthesis

Exam tip   OCR

✔ If you increase a limiting factor then you will increase the rate. If you decrease the factor, you decrease the rate. Remember to be clear about an increase or a decrease, rather than saying 'photosynthesis depends on the factor'.

Questions

1   Describe why plants grow most in spring and summer.   ↓ E

2   Explain why burning a fuel in a glasshouse will increase the rate of photosynthesis.   ↓ C

3   Explain what a limiting factor is.

4   Explain in terms of limiting factors why gardeners do not need to mow their lawns during the winter.   ↓ A*

Learning objectives

After studying this topic, you should be able to:

- ✔ know that the leaf is the site of photosynthesis
- ✔ appreciate the internal and external structure of the leaf
- ✔ understand the adaptations of the leaf for photosynthesis

Key words

leaf, stomata, palisade layer

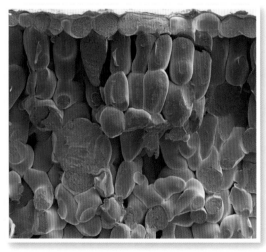

▲ Cross section of a spinach leaf seen through a powerful electron microscope (×340)

A On a plant, leaves are angled so plenty of sunlight reaches them. Explain why this is important to the plant.

B The leaf epidermis is transparent. Why is this an advantage to the leaf?

C What is the name for the pores in the leaf?

Leaves

The main plant organs for making food are the leaves.

▲ A leaf

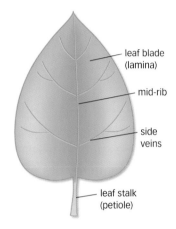

leaf blade (lamina)

mid-rib

side veins

leaf stalk (petiole)

▲ The external structure of a leaf

Inside the leaf

The **leaf** is made of many specialised cells.

> The outer epidermal cells are transparent to allow light through, so contain no chloroplasts

Inside the leaf, the palisade and spongy mesophyll cells are full of chloroplasts. Chloroplasts contain the chlorophyll and other pigments that absorb light energy for photosynthesis. In the lower epidermis, pores called **stomata** are protected by guard cells that open and close to allow gases in and out.

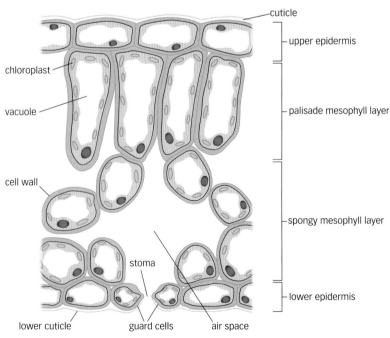

cuticle

upper epidermis

chloroplast

vacuole

cell wall

palisade mesophyll layer

spongy mesophyll layer

stoma

lower epidermis

lower cuticle   guard cells   air space

▲ The internal structure of a leaf

Top ten adaptations of the leaf for photosynthesis

✓ Many leaves are broad and flat, giving a large surface area to absorb as much light as possible.

✓ Leaves are thin, providing a short diffusion pathway for carbon dioxide to diffuse to the mesophyll and palisade cells.

✓ The leaf cells contain chlorophyll and other pigments that absorb energy from different parts of the spectrum.

✓ The cells of the palisade layer are neatly packed in rows, to fit more cells in.

✓ Veins contain vascular bundles. These form a network that supports the leaf blade. They also carry water from the root to the leaf, and carry soluble sugars away.

✓ There are plenty of stomata, pores in the lower epidermis, which allow carbon dioxide in and oxygen out. Guard cells control whether they are open or closed.

✓ The upper **palisade layer**, which receives the most light, contains the most chloroplasts.

✓ There are air spaces in the spongy mesophyll layer to allow carbon dioxide to diffuse from the stomata to the palisade cells.

✓ The air spaces inside the leaf give a large surface area to volume ratio. This allows maximum absorption of gases.

✓ The epidermis is transparent.

How scientists' ideas have developed

The process of photosynthesis is now common knowledge. But scientists haven't always known how photosynthesis works. Their understanding of the process has been slowly built up over time.

The Ancient Greeks thought that plant growth was the result of plants absorbing minerals from the soil. In the 1600s, Belgian scientist Jean-Baptiste Van Helmont grew willow trees, keeping each in the same pot for several years, adding only water. The mass of the willow tree increased by 74 kg during this time, but there was no real change in the mass of the soil in the pot. He concluded that plants grow by absorbing water.

Later still, biologists realised the importance of carbon dioxide for photosynthesis. In 1771, British biologist Joseph Priestly grew mint plants in a sealed chamber containing some mice. The plants produced the oxygen that the mice needed to survive.

More recently, experiments using isotopes have proved that the oxygen released by the plants came from the water, not the carbon dioxide. It is through work like this that scientists gradually increased their understanding of the science of photosynthesis.

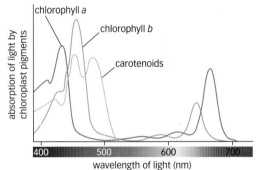

▲ The peaks in this absorption spectrum show the amount of light absorbed by different pigments. Having several chloroplast pigments means that plants can absorb light across a greater range of colour wavelengths. Carotenoids include carotene and xanthophyll.

Questions

1 Which adaptations of the leaf allow it to trap as much sunlight as possible? ↓E

2 What did Van Helmont discover was needed for plant growth?

3 Leaves of plants that are often in bright sunlight tend to have more stomata. Explain what you think the effect of this will be. ↓C

Learning objectives

After studying this topic, you should be able to:

✓ understand the process of diffusion

✓ know how diffusion allows particles to enter and leave cells

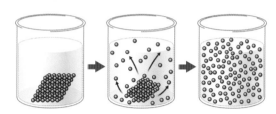

▲ Diffusion is the movement of particles along a concentration gradient

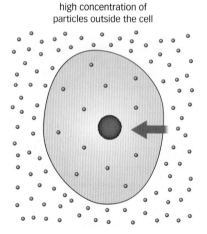

high concentration of particles outside the cell

equal concentration of particles inside and outside the cell

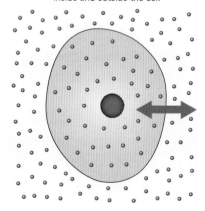

▲ Particles moving into cells by diffusion

Getting in and out

Cells carry out many reactions. They need a constant supply of some substances, and need to get rid of others. So dissolved particles (molecules and ions) need to get into and out of cells. One important way that particles can move into or out of a cell is by **diffusion**.

Diffusion

Particles in a gas or in solution constantly move around. Particles tend to move from an area where they are in high concentration to an area where they are in lower concentration. The particles move until they are evenly spread. This is called diffusion.

Diffusion in cells

Many dissolved substances enter and leave cells by diffusion, including important molecules like oxygen, which is needed for respiration in plant and animal cells. Carbon dioxide also gets into and out of cells by diffusion. Substances can diffuse as gases, or as dissolved particles in solution.

To get into a cell, particles pass through the cell membrane. The membrane will only allow small molecules through. This is fine for oxygen and carbon dioxide as they are both small molecules. The process of diffusion does not use energy, because the molecules move spontaneously from regions of high concentration to regions of low concentration.

Molecules like carbon dioxide and oxygen also diffuse in and out of exchange organs like the leaf in a plant. The leaves are adapted to increase the rate of diffusion. They are large, for example, giving a greater surface area over which gases can diffuse.

> **A** Define diffusion.
> **B** List some important molecules that diffuse into and out of cells by diffusion.
> **C** Explain how the cell membrane can control which substances enter or leave the cell.

Diffusion happens because of constant random movement of particles in solution constantly move. They can move in any direction, but far more particles tend to move from high to low concentration than the other way. This gives a net movement of particles from high concentration to low. However, the rate of diffusion can vary.

Factors that affect the rate of diffusion

Distance

The shorter the distance the particles have to move, the quicker the rate of diffusion will be. For example, if carbon dioxide has to reach cells in the centre of the leaf, then the thinner the leaf, the shorter the distance the gas has to travel and the quicker it will reach the cells.

Concentration gradient

Molecules move from high to low concentration, down a **concentration gradient**. The greater the difference in concentration between two regions, the faster the rate of diffusion. For example, leaf cells produce oxygen as a waste gas during photosynthesis. There is a build-up of oxygen in the leaf, giving a steep concentration gradient of oxygen between the inside and outside of the leaf. This leads to rapid diffusion of oxygen out of the leaf.

Surface area

The greater the surface area that the particles have to diffuse across, the quicker the rate of diffusion. For example, the lungs of animals and the internal structures of a leaf have a large surface area. This allows gases to diffuse rapidly into and out of cells.

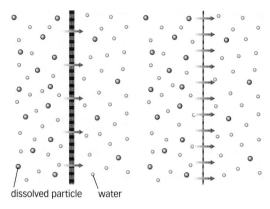

dissolved particle   water

▲ The rate of diffusion depends on the distance the dissolved particles have to travel

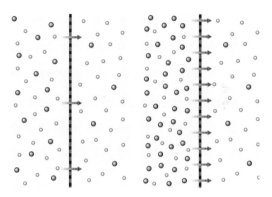

▲ The rate of diffusion depends on the concentration gradient

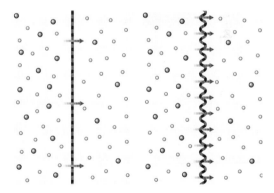

▲ The rate of diffusion depends on the surface area

Questions

1 Does diffusion require energy?

2 For a molecule to diffuse into a cell, should the concentration outside the cell be higher or lower?

3 Oxygen diffuses across the gills of a fish. Do you expect the cells lining the gills to be thick or thin? Give a reason for your choice.

E

A*

Key words

diffusion, concentration gradient

▲ A wilted coleus plant. The cells have lost water by osmosis.

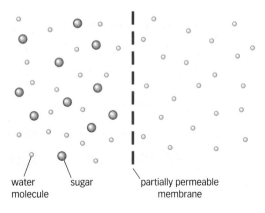

water molecule   sugar   partially permeable membrane

▲ A partially permeable membrane has pores that allow water molecules through, but not larger sugar molecules

Moving water

Osmosis is a special kind of diffusion. Water moves into and out of cells by osmosis.

The cell membrane has tiny holes called **pores**. Larger molecules such as sugars and proteins are too big to pass through the pores, but very small molecules including water can pass through. This type of membrane is called a **partially permeable membrane**, because only some molecules can pass through it.

The diagram on the left shows a dilute sugar solution separated from pure water by a partially permeable membrane. The sugar molecules are too big to pass through the pores. Water molecules pass from the pure water to the sugar solution by diffusion. This dilutes the sugar solution.

In osmosis the is a │ net random   movement of water from an area of high water concentration (pure water or a dilute solution) to an area of low water concentration (a more concentrated solution of sugar or another solute) across a partially permeable membrane.

> **A** In which direction do water molecules move in osmosis?
>
> **B** What is a partially permeable membrane?

Osmosis and cells

Water can move into or out of cells by osmosis. This movement of water is important for both plants and animals, because it keeps their cells in balance.

When plant cells take up water by osmosis, the cells become firm. The cell contents push against the inelastic cell wall. When plant cells are firm like this, it helps support the plant. If they lose water, the cells become soft and the plant wilts.

Osmosis is also important in animals. There is no cell wall in animal cells, so they are very sensitive to water concentrations. If they take in or lose too much water, the cells are damaged and can die.

Osmosis and plant cells

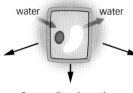
water   water

Surroundings are a less concentrated solution than cell contents (higher water concentration).

Surroundings have the same concentration as cell contents.

Surroundings are a more concentrated solution than cell contents (lower water concentration).

water

Cell placed into a dilute solution. It takes up water by osmosis. The pressure in the cell increases; this is called turgor pressure. The cell becomes firm or turgid.

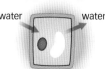
water   water

Cell placed into a solution with the same concentration as its contents. There is no net movement of water. The cell remains the same.

water

Cell placed into more concentrated solution. It loses water by osmosis. The turgor pressure falls. The cell becomes flaccid (soft). Eventually the cell contents collapse away from the cell wall. This is called a plasmolysed cell.

▲ Water movement by osmosis in plant cells

Osmosis and animal cells

water   water

Surroundings are a less concentrated solution than cell contents (higher water concentration).

Surroundings have the same concentration as cell contents.

Surroundings are a more concentrated solution than cell contents (lower water concentration).

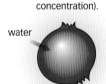
water

Cell placed into a solution that is more dilute than its contents. It takes up water, swells, and may burst. This is called lysis.

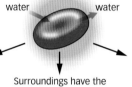
water   water

Cell placed into a solution with the same concentration as its contents. There is no net movement of water. The cell remains the same.

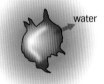
water

Cell placed into a more concentrated solution. It loses water by osmosis. The cell becomes crenated (it crinkles).

▲ Water movement by osmosis in animal cells

◄ A normal red blood cell and a crenated red blood cell

The difference in the ways in which the plant and animal cells respond to this movement of water is due to the plant cell wall.

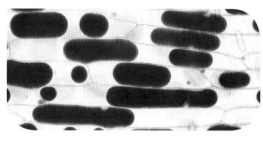
▲ Plasmolysed plant cells

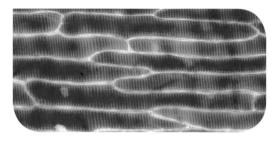
▲ Turgid plant cells

Key words

osmosis, pore, partially permeable membrane

Questions

1 What is the name given to the holes in a partially permeable membrane?

2 A piece of potato is weighed and then placed into a concentrated sugar solution. After 24 hours it is removed, dried and weighed again.

 (a) Describe what would happen to the mass of the potato piece after 24 hours.

 (b) Explain why this has happened.

3 A casualty from a road accident has lost blood. Why are they given a transfusion of blood, not water?

Learning objectives

After studying this topic, you should be able to:

- ✔ know that animal and plant cells are organised into tissues and organs
- ✔ know the main organs of the plant
- ✔ understand the distribution of tissues inside the plant

Plant organs

Organ	Function
Stem	Supports the plant. Transports substances through the plant.
Leaf	Produces food by photosynthesis
Root	Anchors the plant Takes up water and minerals from the soil.
Flower (this is an organ system consisting of three organs: the petal, the stamen, and the carpel)	Reproduction

The stem, root, and leaf are organs. The flower is an organ system.

Organising an organism

In both plants and animals, cells are organised in a very specific way.

- Groups of similar cells work together as a **tissue**.
- Groups of different tissues work together as an **organ**.
- All of the organs build the whole organism.

In plants there are a number of different organs, each with a different function.

> **A** What is an organ?
>
> **B** In an organ system, different organs work together. Why is the flower classed as an organ system?

Inside a plant

Inside a plant organ are tissues made up of similar cells working together. Two major tissues are **xylem** and **phloem** which are found in the vascular bundles.

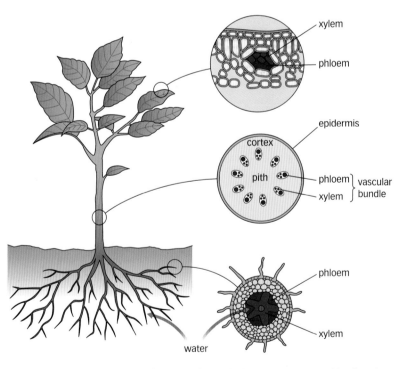

▲ This drawing shows sections cut through the root, stem, and leaf. It shows the different tissues involved in transport around the plant.

A closer look at vascular bundles

The vascular bundles form a continuous transport system from the roots, through the stem, and into the leaves. They carry out two major functions:

- transport
- support.

Structure of xylem and phloem

There are two tissues inside the vascular bundles. Both are involved in the transport of water and dissolved substances through the plant.

Xylem: They are stacked on top of one another to form long hollow tube-like vessels.

These cells are dead and have a hollow cavity called the lumen. Xylem cells are involved in the transport of water and minerals from the roots to the shoots and leaves.

Phloem: They transport the food substances made in the leaf to all other parts of the plant.

These cells are living and are also stacked on top of one another in tubes.

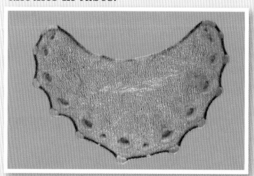

Light micrograph of a section through a celery stem (× 10). A semi-circle of vascular bundles is shown supporting the stem.

Support in plants

The xylem cells have particularly thickened, strengthened cellulose cell walls. These cells help support the plant. The location of the xylem and the vascular bundles helps them carry out their functions.

- In the root the vascular bundles are located in the centre of the root. This helps the root act like an anchor and allows it to bend as the plant moves in the breeze or is tugged from above.
- In the stem the vascular bundles are located around the outer edge of the stem. This provides strength to resist bending of the stem in the breeze.
- In the leaf the vascular bundles form a network which supports the softer leaf tissues.

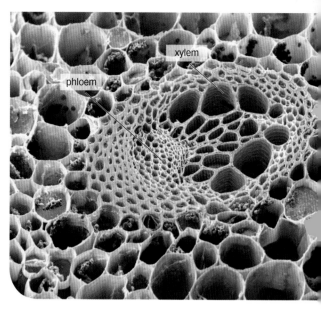

▲ A section through a buttercup stem to show the vascular bundles (× 165)

Key words

tissue, organ, xylem, phloem

Questions

1 Name the tissue responsible for transporting food such as sugars around the plant.

2 Explain why it is important that xylem cells are hollow.

3 Describe how the distribution of the vascular bundles changes at ground level.

4 Explain why the plant needs to transport water to the leaves.

Learning objectives

After studying this topic, you should be able to:

- ✔ know that water and sugars are moved through the plant
- ✔ understand the transpiration stream
- ✔ describe the functioning of the stoma

Moving substances through the plant

Plants can be very big. They need to move substances from one part of the plant to another. They need to move water absorbed in the root, and the sugars made in photosynthesis in the leaves, throughout the plant to the parts that need them. Plants move substances by means of the vascular tissue – the xylem and phloem.

- Xylem continually transports water and minerals up from the root to the leaf. This movement of water is called the **transpiration stream**.
- Phloem transports the sugars made in photosynthesis in the leaf (known as the source) to areas of the plant that are used for storage or are still growing (known as the sink). This process is called **translocation**.

Transpiration

Plants take up water and minerals from the soil by osmosis through their roots. The roots have tiny root hairs, which extend between the soil particles. These greatly increase the surface area of the roots. The water flows up the stem and into the leaf. Water leaves the plant by evaporation and diffusion from inside the leaves, in a process called **transpiration**. The constant flow of water from the roots and out through the leaves is the transpiration stream.

Plants need water. It is important for a number of reasons:

1. Water is needed for the process of photosynthesis.
2. When water evaporates from the leaf it has a cooling effect on the plant and draws water up the xylem.
3. Water enters the cells of the plant by osmosis, and makes the cells turgid or firm. This helps to support the plant.
4. As water moves through the plant, it transports dissolved minerals.
5. The water lost in transpiration needs to be balanced by water uptake from the root.

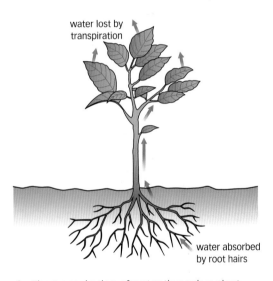

water lost by transpiration

water absorbed by root hairs

▲ The transpiration of water through a plant

A What is the difference between transpiration and translocation?

B Where does water (a) enter the plant (b) leave the plant?

C State two uses of water by the plant.

A closer look at the transpiration stream

◀ Root hairs in a radish plant

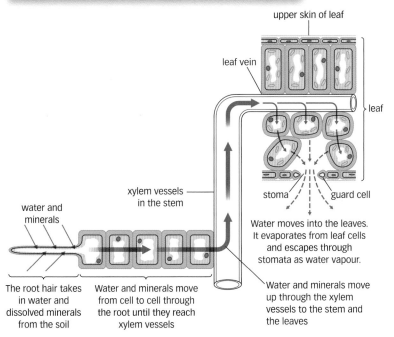

upper skin of leaf

leaf vein

leaf

xylem vessels in the stem

stoma   guard cell

water and minerals

Water moves into the leaves. It evaporates from leaf cells and escapes through stomata as water vapour.

The root hair takes in water and dissolved minerals from the soil

Water and minerals move from cell to cell through the root until they reach xylem vessels

Water and minerals move up through the xylem vessels to the stem and the leaves

▲ The process of transpiration

Controlling water loss

Leaves are highly adapted to be efficient at photosynthesis. A consequence of these adaptations is that the leaves can lose a lot of water by transpiration. To help reduce this, the leaf has a number of mechanisms to reduce water loss:

- A waxy cuticle on the upper and lower surfaces of the leaf does not allow water to evaporate through it.
- Very few stomata on the upper of the leaf.

- Plants that live in dry areas, such as marram grass, often have fewer stomata and they are enclosed on the inner surface of a rolled leaf, protected from the Sun.
- Each stoma can be opened or closed. When the plant is photosynthesising the stomata are open. The stomata are closed at night. When the stomata are closed water loss is reduced.

Guard cells

There are two special cells called guard cells on either side of the stoma. When there is plenty of light and water, the guard cells take up water by osmosis, swell, and become turgid. This causes them to bend and open the stoma. If there is little water, then the guard cells cannot become turgid. Then they do not open the stoma.

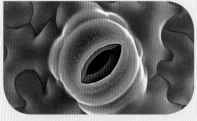

▲ When conditions are good for photosynthesis, the guard cells are turgid, opening the stoma. Carbon dioxide can enter the leaf, water and oxygen can leave.

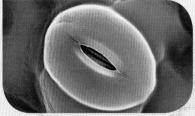

▲ When conditions are not good for photosynthesis, the guard cells close the stoma. This reduces water loss.

Questions

1 Name two ways that the plant reduces water loss.

2 Describe three occasions when osmosis plays a part in the movement of water through the plant.

3 Describe how transpiration helps cause water to be moved up the xylem vessels.

↓ C

Learning objectives

After studying this topic, you should be able to:

✔ know that the rate of transpiration can change

✔ describe and explain how environmental factors can change the rate of transpiration

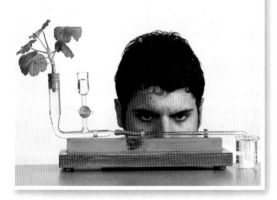

▲ A bubble potometer

A A rate is a speed, which is distance divided by time. What two measurements would you need to take in an experiment using a potometer, to calculate the rate of transpiration?

B Increasing the light intensity will increase the rate of transpiration. How would you notice this using the bubble potometer?

C When comparing the rate of transpiration in two plants, why is it important to conduct the experiments at the same time of day?

Factors affecting the rate of transpiration

There are four main factors in the environment that can affect the rate of evaporation of water. Anything that affects evaporation will affect how quickly water moves through the plant – the **rate of transpiration**. The following factors make the rate of transpiration faster:

- an increase in light intensity
- an increase in temperature
- an increase in air movement
- a decrease in humidity.

Biologists use a piece of apparatus called a bubble **potometer** to measure the rate of transpiration. Using this apparatus, you can change a factor such as the light level, or temperature, and note the change in the rate of transpiration, by measuring how fast a bubble moves along a glass tube. The bubble shows how quickly water is moving through the plant.

Increasing the rate of transpiration
Higher light intensity

Stomata close in the dark and open in the light. When the light intensity is greater, more stomata will open. This allows more water to evaporate, so the rate of transpiration will be faster.

▲ Graph of transpiration rate against light intensity. The rate increases until all the stomata are open, and transpiration is at a maximum.

◀ A higher light intensity increases the rate of transpiration. The stomata open to allow carbon dioxide into the leaves for photosynthesis.

Increase in temperature

The higher the temperature, the faster the particles in the air will move. This means that water molecules move faster and evaporate from the leaf quicker. So the rate of transpiration will increase.

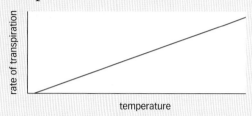

▲ A warmer temperature increases the rate of transpiration

Increased air movement

When air moves over the leaf, it moves evaporated water molecules away from the leaf. The faster the air movement, the quicker the water will be moved. This increases the diffusion of water out of the leaf, because water molecules do not build up in the air outside the leaf. The concentration of water outside the leaf is kept lower, keeping a high concentration gradient between the inside of the leaf and the air outside. So the rate of transpiration increases.

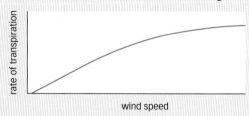

▲ The rate of transpiration is higher on a windy day

Decreased humidity

The less humid the air, the less water there is in it. This again makes for a greater concentration gradient between the inside and outside of the leaf. Water molecules will diffuse out more quickly, so increasing the rate of transpiration.

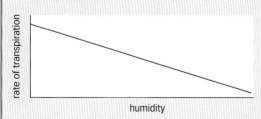

▲ The rate of transpiration is higher when the air is less humid

Exam tip OCR

✔ Try to remember the factors affecting the rate of transpiration by thinking of the best conditions for drying clothes.

Questions

1 Why do gardeners need to water their plants more in the summer?

2 Explain why plants on a sand dune will lose water faster than plants in a woodland.

3 Why do florists spray ferns with water to help keep them healthy?

C

Learning objectives

After studying this topic, you should be able to:

- ✔ know that plants need minerals to maintain healthy growth
- ✔ understand the uses of some plant minerals and the effects of deficiencies in them

▲ Plant fertiliser showing the relative proportions of three important minerals: N (nitrogen), P (phosphorus), and K (potassium)

A List two ways of adding minerals to soil.

B Name the three main minerals in chemical fertilisers.

C State what happens to plants if they do not get enough minerals.

Key words

minerals, deficiency symptom, active transport

Healthy plants

Plants make glucose by photosynthesis. As well as glucose, they also need **minerals** to remain healthy. Plant root hairs absorb small amounts of minerals which are dissolved in the soil water. Once inside the plant, these minerals are used to make useful molecules. Without these minerals plants become unhealthy. They show **deficiency symptoms**. If plants are grown experimentally without soil, but in a solution lacking one mineral, scientists can identify the deficiency caused.

The minerals in the soil water are at low concentrations. Sometimes the mineral levels fall too low. This happens if a particular mineral has been used up. Plants will not grow well if one or more minerals is missing from the soil. Gardeners try to keep their plants healthy by making sure the soil has enough minerals. To do this they can add minerals to the soil by

- adding manure, which decays slowly and releases minerals
- adding compost or rotting leaves that will decay slowly and release minerals
- adding chemical fertilisers, which dissolve and release minerals into the soil.

Which minerals do plants need?

Mineral	Why is it needed?	Deficiency symptom	Result
Nitrogen (N), contained in nitrates	To make amino acids, which are used to build proteins for cell growth.	Poor plant growth; yellow leaves	
Phosphorus (P), contained in phosphates	In respiration, to make an energy-storing molecule (ATP) which is used in growth. Phosphates are also needed to make DNA and molecules in the cell membrane.	Poor root growth; stunted plant; discoloured purple leaves	

Mineral	Why is it needed?	Deficiency symptom	Result
Potassium (K)	Needed for enzymes involved in photosynthesis and respiration. Without it there is not enough food and energy for flowers and fruit to grow.	Poor flower and fruit growth; yellowed leaves with brown spots	
Magnesium (Mg)	Needed to make chlorophyll for photosynthesis.	Yellow leaves, especially the lower leaves	

D Which mineral is missing if plants show purple leaves?

E Explain why magnesium is needed by plants.

Questions

1 Which part of the plant takes in minerals?

2 Suggest why gardeners swap from a high nitrogen feed in the early spring to a high potassium feed in the summer.

3 State two differences between active transport and diffusion.

↓ E

↓ C

↓ A*

F Describe the effect of a lack of nitrogen on plants.

Fertilisers will list the amounts of nitrogen (N), phosphorus (P), and potassium (K) as the NPK ratio. Different fertilisers will have varying amounts of each of these minerals. High nitrogen fertilisers are used to promote leaf growth, while high potassium fertilisers promote flowering.

Active transport

Minerals are usually present in very low concentrations in the soil, lower than their concentration in the plant's cells. Because the cells have a higher concentration, the minerals cannot move into the cells by diffusion. So another method is used to move molecules across the cell membrane and into the cell. This is **active transport**. Key features of active transport are as follows:

* Active transport pumps particles against a concentration gradient (from low to high concentration).
* It requires energy from respiration in the molecule ATP.
* It needs a carrier protein in the membrane.

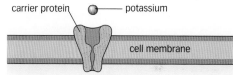

1. A potassium particle attaches to the carrier protein.

carrier protein — potassium

cell membrane

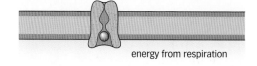

2. The carrier protein uses energy to change shape.

energy from respiration

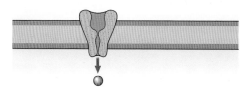

3. The potassium particle moves inside the cell.

▲ A potassium particle moving across a membrane by active transport

Learning objectives

After studying this topic, you should be able to:

- ✔ know that nature recycles by the decay of dead material
- ✔ know that microbes play an important part in the process of decay
- ✔ know the best conditions for decay

Key words

decay, recycling, microbes

▲ Compost is a product of natural recycling

▲ At a sewage works bacteria break down waste

Round and round

Elements pass between the living world and the non-living world – air, water, soil, and rocks – in a constant cycle. Plants absorb elements including nitrogen and carbon, and build them into useful molecules which help the plant to grow. When an animal eats a plant, the plant's molecules become part of the animal.

Eventually all plants and animals die. Their bodies **decay** and this decay process releases the elements back into the environment for the plants to reuse. And so the cycle keeps turning.

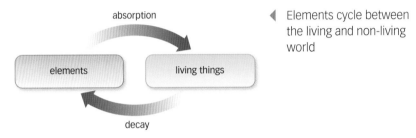

◄ Elements cycle between the living and non-living world

This is a kind of natural **recycling** process. Nature breaks down the remains of plants and animals to return the elements, so they can be used again.

In a natural stable community like a woodland, this cycle keeps turning at a steady rate. The processes that remove materials from the environment and lock them up in plants are balanced by the processes of decay, which return the materials to the environment.

Microbe recyclers

There are two main groups of **microbes** involved in decay: bacteria and fungi. Both can be observed experimentally if bread is left to decay in a lab. Humans make use of these microbes to help us break down waste.

- Compost – gardeners use natural recycling processes. They gather their waste such as grass cuttings, leaves, and twigs. They pile these up or put them into a compost bin, and allow them to decay. The result is a nutrient-rich soil which can be used for growing plants.
- Sewage works – the sewerage produced in homes and factories is sent to a sewage works. Here bacteria digest the organic waste in large tanks, making the sewage clean enough to safely discharge into natural waterways.

Decay happens faster at certain times of the year, when the decay microbes have the right conditions to survive:

Condition	Effect on rate of decay	Explanation
Temperature	At warmer temperatures the rate of decay is faster.	Microbes are able to respire faster, and will grow and reproduce quicker in warmer conditions.
Amount of oxygen	The more oxygen, the faster the rate of decay.	The more oxygen there is available, the faster the microbes will be able to respire. This will allow them to grow quicker and reproduce faster.
Amount of water	In moist conditions decay is faster.	Microbes need water to remain healthy. In moist conditions the microbes will grow faster and reproduce more.

Breaking down the dead

A number of organisms play a role in the process of decay. They break down detritus (dead plants and animals, and animal waste). There are two main groups of decay organisms:

- Detritivores, such as earthworms, maggots, millipedes, and woodlice, eat small parts of the dead material, which they digest and then release as waste. This activity increases the surface area of the dead remains for decomposers to act on.
- Decomposers such as bacteria and fungi chemically break down dead material, releasing ammonium compounds into the soil.

▲ A millipede eating leaf litter – a detritivore

▲ Fungi on dead wood – decomposers

A Use your knowledge of microbes to suggest why the carbon cycle slows during the winter.

B Explain why an increase in (a) oxygen level and (b) temperature will speed up decay.

Saprotrophic feeding

Most decomposers are saprophytes. They feed by releasing enzymes onto the dead animal or plant. The enzymes digest the dead material in a process called extracellular digestion. The decomposers then absorb the digested chemicals. This process is called saprotrophic feeding.

Questions

1 Describe why bacteria are important in natural recycling.

2 Describe how an organic farmer, who does not want to use manufactured fertilisers on their farm, could produce compost to help their crops grow.

3 Describe the difference between a saprophyte and a detritivore.

Learning objectives

After studying this topic, you should be able to:

- ✔ know that food will decompose, and that this is caused by decomposers
- ✔ be aware of various methods to prevent food decay

I'm not eating that!

People have always used different methods of keeping food fresh. At certain times of year food is plentiful. We need to store it so it can be eaten in more difficult times when food is in short supply. We do not just need to prevent food becoming stale – we need to stop the food going off or decaying. Unhelpful microbes are responsible for the decomposing and decay of foods.

▲ Food spoilage is caused by decay microbes

Over the years, people have developed a number of techniques to prevent food decay. Biology can explain how these **preservation** methods prevent food decay.

The cost of decay

There are a number of consequences of food decay:
- It reduces the amount of food for people to eat.
- In some areas of food shortage, decay of food stores could cause malnutrition.
- Eating food that is decaying can lead to illness.
- The profits of farmers and supermarkets are reduced when food decays before it is sold.

A What causes food to decay?

B Suggest why eating food that is going off could make you unwell.

C Why are techniques to preserve food important in a developing country?

Preventing decay

Method	Explanation
Canning	Food is sealed in a metal container. It is heated to kill any microbes. The can prevents the entry of oxygen and any decomposer microbes.
Cooling	Food can be placed in a fridge, typically at 4 °C. At this temperature the reproduction of decomposers is slowed down.
Freezing	Freezers keep food at a lower temperature, often –5 °C. At this temperature decomposers stop reproducing.
Drying	Some foods are dried, such as pasta. Without water decomposers do not grow or reproduce.
Adding salt	Some foods are salted, such as cured meats and fish. Salt causes water to be drawn out of the microbes by osmosis, and this kills them.
Adding sugar	Other foods have sugars added, such as jams. The effect is the same as salting; water is withdrawn.
Adding vinegar	Vinegar is an acid, which is added to foods such as pickled onions during pickling. The acid kills decomposers.

Key words

preservation

Questions

1 Name the seven common ways of preserving food.

2 Explain why food from a tin that has been opened would decay more quickly than a jam.

3 Explain why keeping food in a fridge doesn't stop it from decaying for long.

4 How is preserving food by salting similar to preserving by adding sugar?

A Explain why intensive farming techniques are used.

B Why is intensively reared food cheaper than food produced by traditional farming?

▲ Spraying a crop with insecticide

Food production

Farming produces food for the human population. There are many types of farm. Intensive farms try to produce the maximum amount of food per hectare of land. As the human population has increased, farms have needed to increase their production of food. This has led to the development of intensive farming practices.

Advantages of intensive farming

Advantage	Explanation
High yield, large amounts of food produced	Maximum production is achieved per unit of land. The use of pesticides means that less food is lost to pests.
Low cost of production	The maximum output is achieved from the land available, such as by using fertilisers to increase plant growth.
Less labour intensive	The use of artificial chemicals and machines means fewer people are needed to do the work.

Disadvantages of intensive farming

Disadvantage	Explanation
Pesticides	If not used carefully these may damage the environment.
Fertilisers	If too much soluble fertiliser is used, it can wash into streams and lead to pollution.
Battery rearing of animals in small enclosures	This tends to be less humane, and it can cause disease to spread quickly through the animal population.

Some intensive farming techniques

Two important intensive farming techniques are

- the use of pesticides
- battery farming.

Pesticides

Pesticides are chemicals that kill pests. There are several different types of pesticide:

- Insecticides kill insects, which might be a pest because they eat the crop the farmer is growing, or because they spread disease among animals.

- Fungicides kill fungi, which can lead to the decay of plants. Some fungi can cause illness in animals, such as ringworm in sheep.
- Herbicides kill weeds, which might compete with the crop for resources such as light, water, and soil minerals.

The use of pesticides reduces damage to the crop or herd. The pests are stopped from competing with the crop or herd for resources. There are no weeds competing with the crop for sunlight energy, so the crop has the maximum rate of photosynthesis. More plant material is available to be passed to the next link in the food chain. So, for example, when grass is eaten by a sheep, no energy has been lost to the pest.

At every link in the food chain less energy is lost to pests. This increases the yield for the farmer, and also increases profits.

The disadvantage of pesticides is that they are artificial chemicals that can enter and build up or accumulate in the food chain. Some pesticides can directly affect human health.

A group of insecticides that were used for many years were dioxins (see diagram on right). These are poisons that can build up in the food chain and cause problems for wildlife, because they are persistent and do not break down.

Organophosphates are a group of insecticides that were used in sheep dips. These chemicals killed insects living in the sheep's wool. However, it was discovered that they caused nervous system problems in farmers who used them.

Battery farming

Battery farming is a technique in which large numbers of animals are reared indoors. The advantage to the farmer is that the animals cannot move around as much and are kept warmer, which stops them wasting energy. The disadvantage is that it is less humane, the animals cannot roam freely, and their behaviour changes as a result.

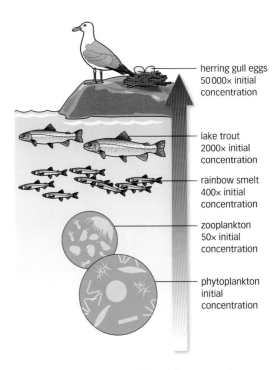

herring gull eggs
50 000× initial concentration

lake trout
2000× initial concentration

rainbow smelt
400× initial concentration

zooplankton
50× initial concentration

phytoplankton
initial concentration

▲ Dioxins were washed into lakes. Here they were absorbed in small amounts by microbes called phytoplankton. The phytoplankton were eaten by zooplankton. Dioxins do not break down in their bodies, so as they eat a large number of phytoplankton, the concentration of the poison builds up in their body. This effect continues, causing toxic levels which poison animals further up the food chain such as water birds. Dioxins have now largely been banned.

Questions

1 What is the advantage to the farmer of using a herbicide? **E**

2 Why are most insecticides required by law to break down quickly in the environment? **C**

3 Some people regard battery farming as inhumane. Discuss the pros and cons of battery farming. **A***

Learning objectives

After studying this topic, you should be able to:

- ✔ know that new farming techniques have been developed to increase yield
- ✔ know the major advantages of the techniques

▲ A salmon farm in Scotland

A Explain why the cost of salmon in the supermarket has decreased in recent years.

B Explain why biologists needed to develop techniques to increase food production.

▲ Growing vegetables commercially in a large glasshouse

Futuristic farming

The need to produce ever-larger yields of food has led biologists to develop many innovative techniques. These practices make better use of the land available to us. They are intensive practices, but they do not have the same environmental impact as some more traditional methods. These techniques may also be more humane to animals.

Three interesting modern farming techniques are

- fish farming
- **glasshouses**
- hydroponics.

Fish farming

Fish is a healthy food option. Many fish such as salmon and trout have become popular in recent years. However, we cannot keep harvesting the wild fish population because they cannot sustain increased levels of fishing.

To meet the increased demand in a sustainable way, **fish farms** have been developed. Fish are bred and reared in large cages in rivers or the sea. Fish farming techniques are on the increase in many places.

Advantages of fish farming

- There is a large captive stock of fish that are easy to catch.
- It is cheaper to rear fish in farms than to fish for wild stocks, and so farmed fish are cheaper to buy.
- There is less predation of the stock.
- There is less need to fish the wild stock, so their numbers can recover.

Disadvantages of fish farming

- Because the fish are kept close together, any diseases will spread quickly, and could escape and infect the wild population.

Glasshouses

Glasshouses for growing crops out of season are not new. Our Victorian ancestors used them extensively. But we are using them differently these days.

Glasshouses are now larger; some are the size of a football pitch. They have energy-efficient methods to control the environment, such as ventilation and watering systems. In glasshouses farmers can grow commercial crops of fruit or vegetables all year round.

Advantages of glasshouses

- Farmers can manipulate the environment and grow tender crops all year round.
- Diseases can be treated and controlled inside the glasshouse more easily than in fields.
- Pesticides and fertilisers are contained inside, so they do not escape into the natural environment.
- Glasshouses can be placed all around the UK, so we do not have to import so many crops from abroad.

> C   Explain why it is an advantage to use glasshouses, rather than import crops from abroad.
>
> D   How can we have UK-grown strawberries out of season?

▲ Strawberry crops being grown by hydroponics

Hydroponics

Perhaps the most futuristic approach to food production is the use of **hydroponics**. Inside glasshouses, plants are grown without soil. They are suspended with their roots exposed, and sprayed with a solution containing the correct concentration of minerals dissolved in the water. Sometimes the roots are bathed in the solution in a bag. The glasshouse may have several rows of the plant stacked above one another. This means that several sets of plants can be grown in the same space. Any mineral solution that is not absorbed by the plant is collected and recycled, reducing waste.

▲ Lettuce and tomato plants being grown by hydroponics

Advantages of hydroponics

- Plants can be grown in areas with poor soil, because no soil is required.

- The technique gives better control of the minerals needed by the plant.
- There is better use of space.
- Water is recycled, reducing waste.
- Diseases can be controlled inside the glasshouse.

Disadvantages of hydroponics

- There is no support for the plant as roots are not anchored in the soil, so a frame or tray is needed.
- Because there is no soil to hold and store minerals, fertilisers need to be added constantly as dissolved minerals.

Questions

1   Name two environmental factors that can be controlled in a glasshouse. ↓ E

2   Why would hydroponics be a useful technique in desert areas? ↓ C

3   Explain how hydroponics makes the maximum use of minerals in solution. ↓ A*

The green farmer

Organic farmers do not use intensive farming methods. High yield is not their main aim; they try to produce smaller amounts of healthy, good quality food. Their approach does not make great use of artificial chemicals. So how does the organic farmer solve the problems for which the intensive farmer uses pesticides and fertilisers?

There are two major challenges for any farmer:
- promoting the growth of crops
- dealing with pests.

Improving plant growth

Plants need a supply of minerals to grow well. Organic farmers need to add minerals to the soil and maximise crop growth without using artificial fertilisers or herbicides. There are a number of organic methods:

- They can use animal manure and compost made from leaf litter. This not only adds the minerals but also improves the fibre content of the soil, so that it retains water better. The mineral release is slow, but it is maintained over a longer period of time as the wastes are slowly decayed. It can sometimes be difficult in practice to obtain and spread enough manure.
- Weeding removes competition from other plants. This is labour intensive. Some farmers cover the soil with plastic sheeting to prevent weeds from growing.
- Crop rotation means planting a field with different crops in successive years, in a cycle. Organic methods include the regular planting of leguminous plants once in a cycle. These plants have nitrogen-fixing bacteria in nodules in their roots, which add nitrates to the soil. The process is slow, and fields gradually become lower in nitrates during each cycle.
- Organic farmers vary their seed planting times, planting in batches. This means that they harvest a small number of plants regularly, and they do not have to preserve them. Varying seed planting times can also avoid crop growth coinciding with an increase in pest numbers.

▲ An organic farmer weeding between lettuces

Dealing with pests

Organic farmers don't use pesticides like an intensive farmer does. Instead they may use a technique called **biological control**. They introduce a natural predator for the pest, which will kill and eat the pest.

🔺 The ladybird is a natural predator of the aphid or greenfly, which is a pest to many plants

Advantages and disadvantages of biological control

Advantages

- There is no need for artificial chemicals to be used.
- There are no chemicals to escape into the environment and damage or kill other animals.
- Chemicals often need to be reused, but biological control does not usually need a repeat treatment in a season if the predators survive for a long period.

Disadvantages

- The predator may not eat the pest.
- The predator may eat other useful species.
- The predator may increase in number and become out of control.
- The predator may not stay in the area where it is needed.
- Adding organisms to or removing them from any food web may have an impact on that web. It will change the numbers of other organisms in the web. This is also a disadvantage of chemical control, which kills large numbers of organisms within a food web.

A Describe how organic farmers fertilise their fields without using chemical fertilisers.

B Other than adding minerals to the soil, what are the advantages of using manure to fertilise the land?

C What is a leguminous plant?

Questions

1. Name two types of chemical that are not used by organic farmers.

2. Describe two ways in which organic farmers control weeds.

3. Describe the ways in which biological control is better than pesticides for controlling pests.

4. Explain two reasons why organic food is more expensive.

5. What would a farmer have to consider before introducing a predator as a form of biological control?

Module summary

Revision checklist

- Scientists use sampling techniques to study biodiversity in ecosystems.
- Biologists use keys to identify the different plants and animals in a community.
- Plants make their own food, during daylight, by photosynthesis. They make glucose that they can change into other chemicals.
- The rate of photosynthesis increases when there is greater light intensity, more carbon dioxide and a warmer temperature.
- Leaves of plants are well adapted for photosynthesis. There are many of them, they are thin and flat, have stomata for gaseous exchange, and special cells with lots of chloroplasts.
- Substances need to pass into and out of cells. Some substances do this by diffusion.
- Water moves into and out of cells by osmosis.
- Cells are organised into tissues; tissues into organs, and organs into an organism. Stems, roots, leaves, and flowers are plant organs. Xylem and phloem are plant tissues.
- Plants lose water from leaves. This is called transpiration. It causes a stream of water, with minerals, to move up from the roots to the leaves. Plants that live in dry places have mechanisms to reduce water loss.
- The rate of transpiration changes with changes in temperature, air movement, light intensity and humidity.
- Plants need minerals for healthy growth. Adding manure or fertilisers to soil increases the minerals.
- The amount of minerals on Earth is finite, so they have to be recycled. Microorganisms help decompose dead matter and this releases minerals for re-use.
- If food decays, it is not fit to eat, so scientists have developed ways of preserving food.
- Farming produces food for humans. Many farms are intensive and grow the maximum amount of food per hectare.
- New types of farming will be needed in the future, to produce more food for the world. This will include fish farming, glasshouses, and hydroponics.
- Organic farming does not use intensive methods. It produces lower yields and uses more labour, so its products are more expensive.

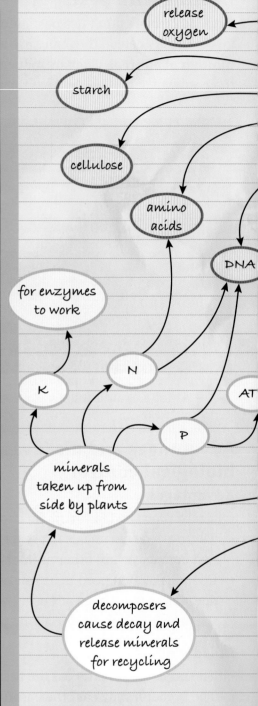

release oxygen

starch

cellulose

amino acids

DNA

for enzymes to work

N

K

AT

P

minerals taken up from side by plants

decomposers cause decay and release minerals for recycling

NOW USE THE B4 GRADE CHECKER ON PAGE 246

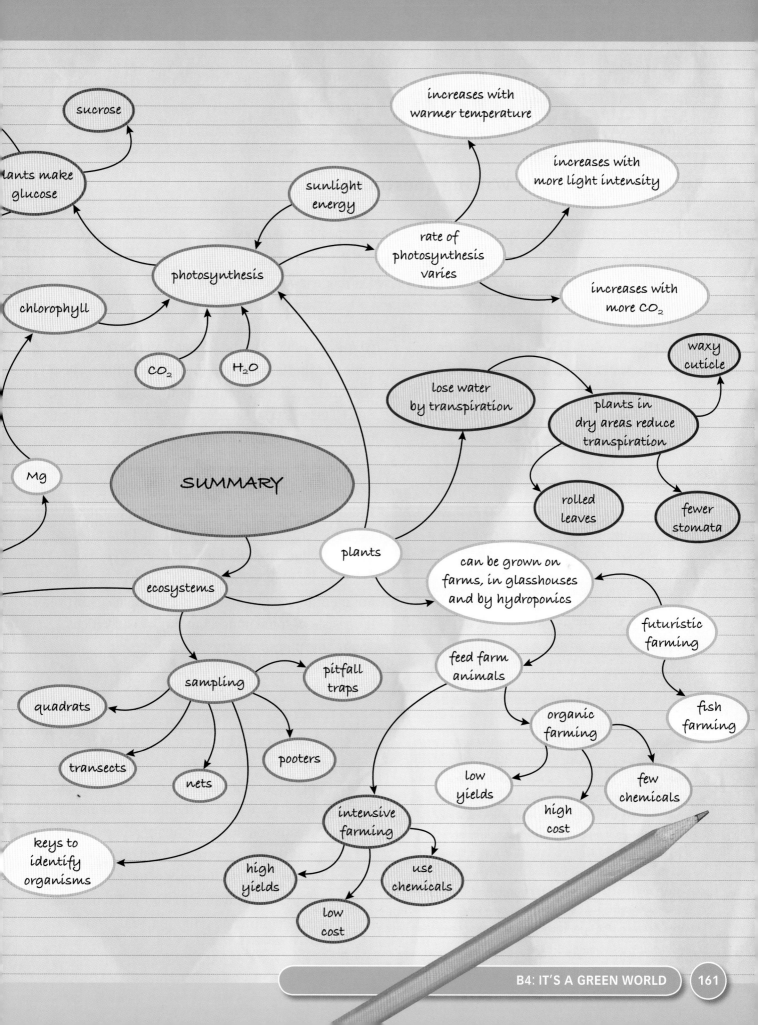

sucrose

plants make glucose

sunlight energy

increases with warmer temperature

increases with more light intensity

rate of photosynthesis varies

increases with more CO₂

photosynthesis

chlorophyll

CO₂

H₂O

waxy cuticle

lose water by transpiration

plants in dry areas reduce transpiration

Mg

SUMMARY

rolled leaves

fewer stomata

plants

can be grown on farms, in glasshouses and by hydroponics

ecosystems

futuristic farming

sampling

pitfall traps

feed farm animals

quadrats

fish farming

organic farming

transects

pooters

low yields

few chemicals

nets

high cost

intensive farming

keys to identify organisms

high yields

use chemicals

low cost

OCR gateway *Upgrade*

Answering Extended Writing questions

QUESTION

Describe how scientists have come to understand the process of photosynthesis.

Refer to the views of scientists in ancient Greece as well of those of more modern scientists.

The quality of written communication will be assessed in your answer to this question.

Plants make food using carbon dioxide from air, water and energy from the sun. They make sugar in the leaves. Their has to be chorofill. Plants give off oxygen which animals breathe in.

↓ E

Examiner: The candidate has described photosynthesis, although the answer should specify that energy comes from sunlight, not just from 'the Sun'. There is no reference to how scientists' ideas have developed. There are some spelling mistakes.

Scientists used to think that plants got there food from soil. Then someone grew a tree in a pot and it gained a lot of wait but not much soil was used up. A mouse in a jar with a plant lived longer because the plant gave off oxygen. Plants use sunlight, water and carbon dioxide to make glucose and oxygen.

↓ C

Examiner: No scientists' names are mentioned in this answer. There are also some spelling mistakes. The candidate understands what scientists now know about photosynthesis, although chlorophyll is not mentioned. The answer is a bit vague. However, it does show some understanding of early ideas about how plants obtained food.

Photosynthesis took years to explain. Ancient Greek scientists like Artistotle, thought plants take in minerals from the soil. In the 1600's van Helmont grew a willow and found that the plant gained more mass that was lost from the soil. The only thing he added was water. This shows minerals and water were needed for growth. Priestley also studied photosynthesis and showed that plants release oxygen. We now know from work done using radioisotopes that the oxygen came from the water, which was split inside the chloroplast using light energy from the sun. The hydrogen from the water combined with carbon dioxide to make glucose.

↓ A*

Examiner: This question is about how science works. The contributions of three important scientists are described here in a logical sequence. Although van Helmont and Priestly did not know about photosynthesis, what they discovered helped other scientists to understand this process in plants. This answer is accurate and well written, with good spelling, punctuation, and grammar.

Exam-style questions

1 Gardeners place organic matter into compost bins to decay and form compost.

A01 **a** Which organisms cause the decay?

viruses   bacteria   birds

A01 **b** Why is the temperature at the centre of the compost high?

  i heat is produced by leaves carrying out photosynthesis

  ii heat is released when microorganisms respire.

A01 **c** How does adding compost to soil aid plant growth?

A01 **d** Holes allow gases in and out.

  i Which gas will enter and why?

  ii Which gas may leave?

2 Young tuna fish are caught by fish farmers and reared in large pens in the sea. They are sold when they reach a mass of 400 kg. The graph shows the effect of feeding different diets.

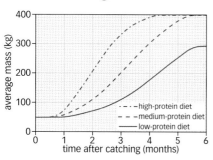

A02 **a** Calculate the mean increase in mass per month of the fish fed on the medium protein diet for six months.

A02 **b** What is the advantage of the high protein diet over the medium protein diet?

A02 **c** What other information does the farmer need to decide whether to use high or medium protein food?

A02 **d** Some consumers will not buy tuna grown in this way. Suggest why.

3 In an investigation potato chips were weighed before and after being placed in salt solutions for an hour.

salt concentration/M	0.0	0.2	0.4	1.0	2.0
mass at start/g	2.5	2.5	2.6	2.5	2.7
mass at end/g	2.8	2.7	2.7	2.3	2.2
% change in mass	+12.0	+8.0		−8.0	−18.5

A01 **a** By what process do cells in the chips gain or lose water?

A02 **b** How could the result above be made more reliable?

A02 **c** Name two factors that should be kept the same in this experiment to make it valid (fair).

A02 **d** Fill in the missing value.

A02 **e** Why are the changes in mass expressed as a percentage change?

A02 **f** Describe how you could find out the strength of salt solution that causes no change to the mass of the chip.

Extended Writing

4 Describe how you would find out which **A01** plants and animals are present on a school playing field.

5 Explain how leaves are well adapted for **A01** photosynthesis.

6 Describe how environmental factors **A01** change the rate of transpiration in plants. How are plants that live in dry places adapted to reduce water loss by transpiration?

A01 Recall the science

A02 Apply your knowledge

A03 Evaluate and analyse the evidence

B5

The living body

Why study this module?

In this module you will find out about the changes in your body as you grow up and change from a child into an adult. You will find out about menstruation and fertility, and how fertility can be controlled or assisted. You will learn about the stages of human life, such as adolescence. Today adolescence is extended, as many young people want a university education so are not ready to begin a fully independent life until they are in their mid twenties.

However, more of us are living longer. 'Spare-part surgery' is increasingly possible, and you will learn about organ transplants and mechanical devices that can help heart or kidney function.

You will also learn about some of your body systems and life processes, such as your skeleton, digestive system, respiratory system, circulatory system, and excretory processes.

You should remember

1 You are made of cells organised into tissues, organs, and systems – such as the reproductive system, circulatory system, and respiratory system.

2 Your joints and antagonistic muscles enable you to move.

3 The male and female reproductive systems.

4 Puberty and the menstrual cycle.

In 2009, UK Biobank was launched. This is a major UK medical research initiative around 500 000 UK citizens aged 40–69 are taking part. They give blood, saliva, and urine samples that will be frozen and stored. Their DNA will be analysed. The participants also take tests to assess their mental faculties and answer questions about their lifestyle. As time passes, data about the participants' illnesses will be correlated with their genetic and biological data. Hopefully this information, along with the data from the Human Genome Project, will inform medical scientists about how our genes and lifestyle influence our health and life expectancy. This will help produce better treatments and medicines for people in the future. Within your lifetime, it will probably be standard medical practice for everyone to be able to have their genome sequenced.

Key words

cartilage, bone

A Name three functions of skeletons.

▲ The great white shark, *Carcharodon carcharias.* Its skeleton is made of cartilage.

Exam tip | OCR

✔ Remember that bone is living tissue. It needs a blood supply to bring oxygen and nutrients to the cells for their respiration.

The functions of skeletons

Skeletons provide

- support
- protection
- a framework for muscle attachment to allow movement.

Types of skeleton

Some animals do not have a hard skeleton

Many invertebrate animals do not have a skeleton. Those that live in water can become quite large because the water buoys them up. Earthworms are supported by the pressure of the fluid inside their body. The fluid presses outwards against their muscular body wall.

Insects have an external skeleton

Insects, along with spiders and crustaceans (lobsters, crabs, and prawns), have an external skeleton made of chitin. This gives a protective outer covering which supports the animal. It also gives a framework for muscle attachment. These animals all have jointed legs and their skeleton, muscles, and joints allow them to move. However, this hard covering can restrict growth. The animals have to shed their old skeleton at intervals and grow before the new skeleton hardens.

Some animals have an internal skeleton

Fish, amphibians, reptiles, birds, and mammals, including humans, have an internal skeleton. In some fish such as sharks, dogfish, rays, and skates, the skeleton is made solely of **cartilage** (gristle).

Your skeleton is made of **bone**, but there are places where there is still cartilage, such as:

- the tip of your nose
- your outer ears
- at the ends of your long bones, such as limb bones and ribs.

Bone and cartilage are both living tissues. They have blood vessels and nerves and they can grow with the body. The internal skeleton forms a framework. Joints and muscles allow it to move. The many small bones of the spine give great flexibility.

More about bone

While you were growing in the womb, your skeleton was first made of cartilage, which is mainly protein. From about 6 weeks, minerals are deposited into the cartilage. These are mainly calcium phosphate. The cartilage becomes ossified – is turned into bone. Children have more cartilage at the ends of their bones than adults, because they are still growing. Forensic scientists can tell the age of a person from the skeleton, according to how much cartilage is still present. Both cartilage and bone can be infected by pathogens.

Paramedics have to be careful not to move a person who has a suspected bone fracture. Moving someone with a broken backbone could injure their spinal cord.

> **B** Why do children need to drink milk or eat cheese to make their bones strong?

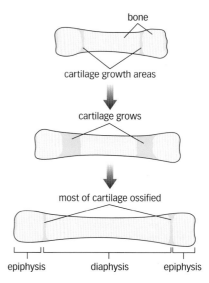

▲ How a bone gets bigger. Cartilage and bone tissue can both grow. Bone also has good powers of regeneration and healing.

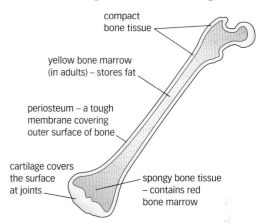

▲ The structure of a long bone

The structure of a long bone

You can see from the diagram that the head of the long bone is covered with smooth cartilage. The outer part of the shaft is hardened bone. It can withstand compression. The shaft is fairly hollow. This makes the bone lighter than solid bones, but still strong. In the centre of the shaft are the bone marrow and blood vessels. Some fat is stored here, and new blood cells are made.

Bones can break

Bones are very strong, but if you knock them they can fracture (break). There are different types of fracture:
- Green stick – the bone is bent but not broken. Children with rickets are prone to this.
- Simple fracture – the bone is broken but the skin is intact.
- Compound fracture – also called an open fracture; the broken ends of the bone stick out through the skin.

Doctors use X-rays to look at the damage done to a broken bone before treating it.

Older people who have osteoporosis (soft bones) are more susceptible to fractures, which may happen in a fall.

Questions

1. List and describe three types of bone fracture. ↓ E

2. Find out how long a broken limb has to be immobilised in plaster. ↓ C

3. Explain how the embryo skeleton changes from being cartilage into bone.

4. Why do you think there is a higher incidence of osteoporosis in Scandinavia? ↓ A*

Learning objectives

After studying this topic, you should be able to:

✔ understand how joints and muscles allow bones to move

✔ identify the main bones of the arm

✔ know that the biceps and triceps work antagonistically to bend or straighten the arm

Key words

tendon, synovial joint, ligament

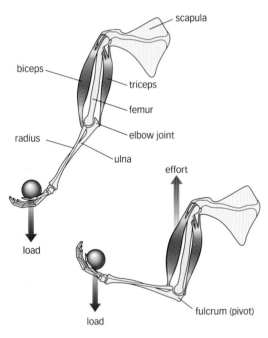

▲ How the arm works as a lever

A Explain how your arm muscles allow you to bend and straighten your arm.

B What are antagonistic muscles?

The human arm

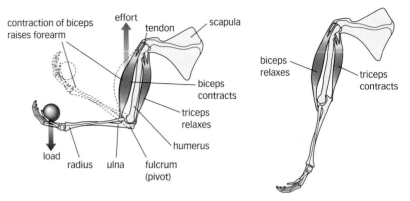

▲ The main bones and muscles of the human arm when bent and extended

When you bend your arm:

• Your biceps muscle contracts and your triceps muscle relaxes.

• As your biceps contracts, the **tendon** that joins it to the radius does not stretch.

• So it pulls the radius upwards and your arm bends.

When you straighten your arm:

• Your triceps muscle contracts and your biceps relaxes.

• The tendon from the triceps pulls on the ulna.

• Your arm straightens.

When a pair of muscles acts together in this way, one contracting and the other relaxing, we say they are antagonistic.

Levers

When the arm bends and straightens it acts as a lever:

• The elbow is the pivot point (fulcrum).

• The hand moves through a larger distance than the muscles.

• The muscles exert a larger force than the load that the hand lifts.

Joints

The contracting muscles provide the force to move your bones. However, you could not move if you did not have joints in your skeleton. Joints are where the end of one bone meets another bone. There are different types of joint.

Fixed joints

Your skull is made of many bones and they join together by fixed joints. At birth, babies have a cartilage patch on top of their skull, so that the bones of the skull can be squeezed as the baby passes down the birth canal. When these joints in the skull are fused, the skull protects the brain well.

Synovial joints

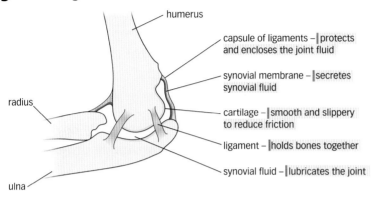

▲ A **synovial joint**

- humerus
- capsule of ligaments – protects and encloses the joint fluid
- synovial membrane – secretes synovial fluid
- radius
- cartilage – smooth and slippery to reduce friction
- ligament – holds bones together
- synovial fluid – lubricates the joint
- ulna

Joints that are freely movable are also called synovial joints. Synovial joints are well adapted to allow smooth, almost friction-free movement:
- The ends of the two articulating (moving) bones are covered in smooth, slippery cartilage.
- The whole joint is enclosed in a capsule.
- Lining the inside of the capsule is a synovial membrane.
- This membrane secretes (makes) synovial fluid.
- Synovial fluid lubricates the joint.

Ligaments join the two bones of a synovial joint together. They stretch and allow movement.

Examples of synovial joints are:

Hinge joints

At your elbow you have a hinge joint. It allows movement in one plane. This allows you to bend and straighten your arm and to also lift heavy weights. Your knee joint is also a hinge joint. You can bend and straighten your leg to walk, but it can also lock into place to bear your weight.

Ball and socket joints

Shoulder and hip joints are ball and socket joints. At these joints you can rotate the limb bones.

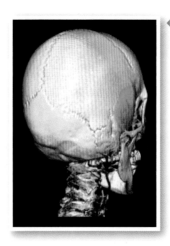

◄ Human skull showing the seams between the different bones

Questions

1 Where do you have fixed joints?
2 Name two places in your body where you have hinge joints.
3 Name two places in your body where you have ball and socket joints.
4 Explain why tendons are not stretchy.
5 What are synovial joints?
6 Describe the functions of the following structures in a joint: (a) synovial membrane (b) synovial fluid (c) cartilage (d) ligaments.
7 When you lift a heavy book, your biceps muscle exerts more force than the weight of the book. However, it enables you to lift the book through a large distance. Explain how it does this.

E

C

A*

3: The circulatory system

Learning objectives

After studying this topic, you should be able to:

- ✔ know that some animals do not have a circulatory system
- ✔ know that some animals have an open and some have a closed circulatory system
- ✔ know that some animals have a single closed system and others have a double closed system

Key words

blood vessels, heart

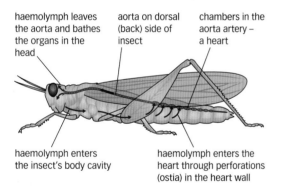

haemolymph leaves the aorta and bathes the organs in the head

aorta on dorsal (back) side of insect

chambers in the aorta artery – a heart

haemolymph enters the insect's body cavity

haemolymph enters the heart through perforations (ostia) in the heart wall

▲ An insect's circulatory system. It is an open system because the blood is not always in vessels.

A Explain why single-celled organisms do not need a circulatory system.

B Why is an insect circulatory system described as an open system?

Single-celled organisms

Small organisms such as amoebae do not need a circulatory system. They have a large surface area compared with their volume. They are surrounded by the water they live in. Dissolved oxygen diffuses from this water into the cell through the cell membrane. Waste material can diffuse out of the cell. There is no need for a special transport system.

Larger animals need a circulatory system because diffusion alone cannot efficiently transport substances to and from their cells. They need blood to do this.

Open systems: insects

Insects have an open circulatory system. They do not have arteries or veins, but their blood flows freely through their body cavity.

The blood makes direct contact with the organs and tissues. Blood travels in the aorta from the heart up to the head. It bathes the organs and muscles of the head and then trickles back through the body cavity. It passes over the gut and gets back into the heart through small holes.

Insect blood does not carry oxygen, because the insect's breathing tubes deliver oxygen directly to the tissues. Their watery, greenish yellow blood carries amino acids, sugars, and ions. There are some white cells to ingest pathogens.

Closed systems: vertebrates

Vertebrates have a closed circulatory system. The blood is contained in **blood vessels** – arteries, veins, and capillaries.

Fish

The blood circulates once around the body, from the **heart** to

- the gills, where it collects oxygen and unloads carbon dioxide
- the body organs and tissues. This is a single circulatory system.

The fish heart is a single pump consisting of two chambers.

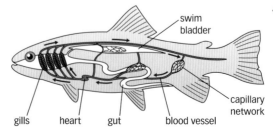

swim bladder

◄ The circulatory system in a fish. Because the blood passes through the heart only once on each circuit around the body, the heart needs only two chambers.

gills   heart   gut   blood vessel

capillary network

Humans

Humans and other mammals have a double circulatory system. There are two circuits from the heart.

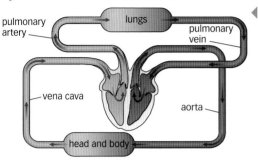

◀ The double circulatory system of humans

Labels: pulmonary artery, lungs, pulmonary vein, vena cava, aorta, head and body

Blood passes

- from the heart to the body organs and tissues
- back to the heart
- to the lungs to remove carbon dioxide and collect oxygen
- back to the heart before being pumped out to the body again.

Because the blood makes two circuits from the heart, the heart needs four chambers. It is a double pump. Blood in a double circulatory system is under higher pressure than blood in a single circulatory system so it transports materials more quickly around the body.

Galen and Harvey

In the second century AD, Galen, a Roman physician (doctor) of Greek origin, dissected monkeys and pigs. He noticed that blood in veins was darker than blood in arteries. He mistakenly thought that the liver made blood and pumped it in the veins to the organs, which consumed it. Galen's ideas influenced medicine for well over 1000 years. In the sixteenth century, Leonardo da Vinci studied anatomy and made drawings showing how blood passed through the heart chambers, and how the heart valves worked.

In the seventeenth century, William Harvey, a doctor at St Bartholomew's hospital in London, published a book showing how blood circulates in the body, from the heart and back again, and to and from the lungs. He realised that the pulse in arteries was linked to the contraction of the left ventricle of the heart. He discovered that veins have valves to prevent backflow. He postulated that there were capillaries, but did not see them as he did not have a microscope.

Exam tip OCR

✔ Remember the difference between a closed circulatory system, where the blood is always in vessels, and an open system, where the blood flows through the body cavity.

Questions

1 Fish and humans both have a closed circulatory system. What is a closed circulatory system? E

2 Why do you think insects do not have red blood cells or haemoglobin in their blood? C

3 Explain why fish need only a two-chambered heart.

4 Why do humans and other mammals need a four-chambered heart?

5 Find out more about the contribution William Harvey made to our understanding of the circulatory system. A*

A Explain why the left ventricle wall is thicker than the right ventricle wall.

B A measurement of blood pressure inside the arteries when the heart beats gives 115 mmHg. In the capillaries it is 40 mmHg, and in the large veins 2 mmHg. Explain these different pressures.

The human heart

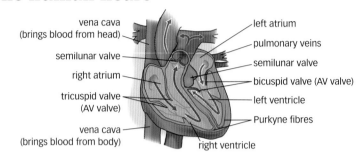

▲ Section through a human heart showing the path of oxygenated and deoxygenated blood

Heart muscles contract to cause the blood to move. It contracts rhythmically to squeeze blood out of the **atria**, into the **ventricles** and then out into the arteries and around the body. The powerful heart muscle needs a continuous supply of glucose, fatty acids, and oxygen so that the muscle cells can respire aerobically and release energy for contraction. The strong contractions of the heart mean that blood leaving the heart in the arteries is at high pressure. As it passes along the circulatory system through increasingly branching arterioles to the capillaries, the blood pressure falls. In the wide tubes of the veins the blood pressure is very low; valves are needed to keep the blood flowing in one direction.

The left ventricle wall is thick muscle that produces enough pressure to push blood in the arteries all over the body. The right ventricle wall is thinner and the muscle is less powerful – it only has to send blood as far as the lungs. The lungs are delicate and too much pressure in the blood entering them could damage them.

The rate at which the heart contracts is controlled by a group of cells called the pacemakers. These produce a small electrical current that stimulates the cardiac muscle to contract.

The pulse: a measure of heart rate

As the ventricles contract and send blood into the arteries, the thick, muscular, and elastic walls of the arteries expand and recoil as a spurt of blood enters. This is the **pulse** that is transmitted all along the length of the arteries in the body. You can detect it where an artery passes over a bone or is near to your skin. Pulse rate is a measure of heart rate.

The cardiac cycle (heartbeat)

The series of events during one contraction is the cardiac cycle.

- The sinoatrial node (**SAN**) produces electrical impulses.
- These spread quickly across the two atria, which contract.
- This forces open the atrioventricular (AV) valves and pushes blood into the ventricles.
- A patch of muscle fibres called the atrioventricular node (**AVN**) conducts the impulses to special conducting muscle fibres, called Purkyne fibres, which carry the impulses to the tip of the ventricles.
- The two ventricles contract. This closes the AV valves and pushes blood out of the ventricles, through the open semilunar valves, into the arteries.
- The atria relax and fill with blood.

Then the cycle starts again. Each cycle is one heartbeat.

Changing the heart rate

When you exercise, your skeletal muscles respire more oxygen and glucose and make more carbon dioxide, which enters the blood. Part of the brain detects the extra carbon dioxide in the blood. It sends impulses to the heart's pacemaker to speed up the heart rate.

You also make more adrenaline when you exercise, and when you are frightened or excited. Adrenaline travels in your blood and affects many target tissues, including the heart's pacemaker, which speeds up the heart rate.

An increased heart rate also delivers more oxygen and glucose to the heart muscle itself, via the coronary arteries. The heart muscle needs to respire more if it is beating more times each minute.

Monitoring the heart

Doctors and cardiac technicians can measure the electrical activity in your heart. They get a trace called an **ECG** (electrocardiogram) and this tells them whether your heart is normal or not. A patient with an irregular heart rate or one that is too fast or too slow may need an artificial pacemaker fitted.

An echocardiogram uses ultrasound to make a scan and show any heart defects.

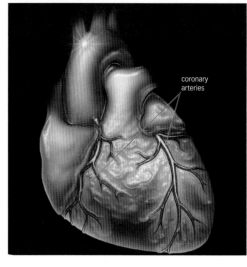

coronary arteries

▲ The coronary arteries, which carry oxygenated blood from the aorta to the heart muscle

a
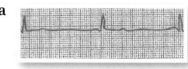

b
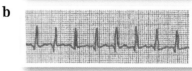

c
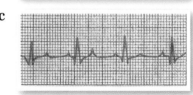

▲ ECG trace of: a, abnormal heart rate – too slow
b, abnormal heart rate – too fast
c, normal heart rate

Questions

1. Explain how heart muscle causes blood to move. **E**

2. Explain why your heart rate needs to increase when you exercise. **C**

3. Find out why some people need an artificial pacemaker.

4. A person exercising has a pulse rate of 120 beats per minute. How long does each cardiac cycle last? **A***

Learning objectives

After studying this topic, you should be able to:

- ✔ know that artificial pacemakers are used to control heartbeat
- ✔ explain the effects of different problems with the heart
- ✔ compare artificial pacemakers and valve replacements with a heart transplant
- ✔ discuss the carrying of donor cards

A Explain why some people need an artificial pacemaker.

B Explain why a patient with a hole in the heart feels tired.

There are many heart diseases and conditions, including:

Artificial pacemakers

Some people have an irregular heartbeat. They can have an artificial pacemaker. This is usually implanted just under the skin in the chest. A wire passes from it into a vein and into the right atrium.

The pacemaker has a long-life battery. It sends impulses to the heart muscle, to make it contract at the correct rhythm. A pacemaker can detect when the person is more active, and sends impulses to the heart at an increased rate.

Some people have a damaged AVN. Its impulses do not travel to the ventricles. In this case the artificial pacemaker wire goes to the ventricle and makes it contract.

Pacemakers have to be replaced about every ten years.

Hole in the heart

Some people have a 'hole in the heart'. Blood can flow directly from one side of the heart to the other. There is less oxygen in the blood, causing fatigue and breathlessness. The hole may be repaired with surgery.

Why a hole in the heart means less oxygen

Fetuses have a small hole between the left and right atria. They obtain oxygen from the placenta. Oxygenated blood enters the right side of the heart and flows through the hole to the left atrium, and then to the head and body. There is a connecting vessel between the pulmonary artery and the aorta, so most blood bypasses the lungs.

At birth, the baby needs to oxygenate its blood via the lungs rather than the placenta. The hole closes and the blood follows the normal circulation. In some babies, the hole does not close, allowing oxygenated blood to flow from the left to the right atrium. This means that a smaller amount of oxygenated blood leaves the left ventricle, and the body tissues receive less oxygen. Patients suffer fatigue and breathlessness. The right side of the heart has to work harder to cope with the increased flow of blood through it. A hole in the heart can be repaired with surgery.

▲ An artificial heart valve. The wires are used to secure the valve in place in the patient's heart. Heart-assist devices may be used after surgery to take the strain off the heart.

Damaged heart valves

As people age, their heart valves may become stiff. Valves may also be damaged, such as by bacterial infection (endocarditis). If the valves in the heart do not close properly, blood will flow backwards. This leads to heart failure, and not enough oxygenated blood can reach body tissues.

Surgeons can replace faulty heart valves with artificial valves or valves from pigs or cows. Because the heart valves have no capillaries supplying them, there is no **rejection** of these transplanted valves.

Blocked coronary arteries

As you get older, and especially if you eat too much saturated fat and smoke tobacco, fatty deposits or plaques build up in your artery walls. These can become quite large and obstruct the flow of blood. If this happens in your coronary arteries, your heart muscle does not get enough oxygenated blood; this is called coronary artery disease. Heart muscle cells cannot respire anaerobically, and without enough oxygen they cannot release enough energy to contract efficiently. You may develop angina, or have a heart attack.

Surgeons can correct this condition with **bypass surgery**. A piece of blood vessel, usually a vein, is taken from the patient's arm or leg and transplanted to bypass the blockage (or blockages) in the coronary artery.

When people have had a heart attack, surgeons can quickly insert a stent, a tube to open up the blocked coronary arteries. If this is done soon enough, the damage to the heart from the heart attack is slight.

Heart transplants

Since the first heart transplant in 1967, many of these operations have been carried out in the UK each year.

A heart transplant is a **traumatic** operation and the recipient must take drugs to suppress the immune system and prevent rejection. There is a shortage of donor hearts. Heart valve replacement and artificial pacemakers are less traumatic with no risk of rejection, but pacemakers and heart valves have to be replaced. All operations carry the risk of infection.

Donor cards

Some people carry donor cards so that doctors can take their organs when they die, without expecting bereaved relatives to make a painful decision. Because there is a shortage of donors, another system could be used. Everyone would be a potential donor unless they carried a card to opt out. Some people would opt out for religious reasons; Jehovah's Witnesses regard blood as sacred.

Key words

bypass surgery

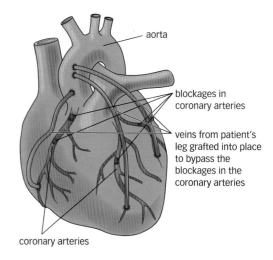

aorta

blockages in coronary arteries

veins from patient's leg grafted into place to bypass the blockages in the coronary arteries

coronary arteries

▲ How blocked coronary arteries are bypassed. The resulting ease of blood flow relieves angina and reduces the risk of a heart attack.

Questions

1. List four diseases or conditions of the heart.

2. What causes a heart attack?

3. Why are there no problems with rejection of transplanted heart valves?

4. Discuss the pros and cons of the opt-out rather than opt-in system for organ donation.

5. Discuss the relative advantages and disadvantages of heart transplant surgery compared with valve replacement or having an artificial pacemaker fitted.

Learning objectives

After studying this topic, you should be able to:

- ✔ describe the reasons for and process of blood donation and transfusion
- ✔ explain the basis of the ABO blood grouping system
- ✔ describe the process of blood clotting

Key words

platelets, fibrin, **agglutinins**

▲ Donated blood is stored in plastic bags. A chemical to prevent clotting is added. The bags are carefully labelled and refrigerated.

Did you know...?

This is how the blood clots:
- The **platelets** in blood are exposed to air.
- This causes a series of chemical reactions.
- Eventually a soluble protein in your plasma changes to insoluble **fibrin**.
- Fibrin fibres form a clot.

Blood transfusions

- In 1818 Dr James Blundell performed the first successful transplant of human blood to a patient with a haemorrhage (bleeding).
- In 1840 a haemophiliac was treated by a blood transfusion at St George's Hospital, London. In haemophilia the blood does not clot, so sufferers who cut themselves keep on bleeding.
- In 1901 an Austrian doctor discovered human blood groups and transfusions became safer. In the next ten years, scientists found that if they added anticoagulant and refrigerated the blood, it would keep for some days.
- Blood banks were established during the First World War.
- In 1950 plastic bags replaced breakable glass bottles for storing the blood.
- People may voluntarily give blood three times a year. This keeps the blood banks full to supply hospitals. Blood is warmed before being transfused into a patient.

In a transfusion, blood groups are now carefully matched between the donor and recipient. There are four main blood groups called A, B, AB, and O. These are further subdivided into groups called Rhesus positive and negative.

Why are anticoagulants needed?

If you cut yourself, your blood should clot to heal the wound. This prevents bacteria from entering and stops blood loss.

Blood kept in a bag would clot in this way. Chemicals are added to block the chemical reactions and prevent clotting (coagulation).

We need vitamin K to help blood clot. Bacteria in our gut make vitamin K, but we can also get it from green vegetables and cranberries.

Abnormal clotting

People with haemophilia have blood that clots very slowly.

People with fatty deposits in their arteries may develop clots in the arteries. These could cause heart attacks or strokes. Smoking tobacco and drinking alcohol increase the risk of blood clots forming. Warfarin, aspirin, and heparin reduce the ability of the blood to clot and can be used to reduce the risk of strokes in some people.

Who needs a transfusion?

Some examples are:
- people who have lost a lot of blood through injury or during surgery
- haemophiliacs
- some cancer patients.

A Why is it useful for your blood to clot when you cut yourself?

B Why are blood clots in blood vessels dangerous?

Why are some blood transfusions unsuccessful?

If the donor and recipient bloods are not matched properly, agglutination or clumping happens. The blood cannot circulate and the recipient dies.

Your red blood cells have proteins called **agglutinins** (a type of antigen) on their membranes. There are different shaped agglutinins. Your blood plasma has antibodies against the agglutinins. You don't have antibodies in your plasma that can react with the antigens on your own red blood cells.

When transfusing blood, doctors have to think about the antibodies clumping the red cells together. They need to consider how the antibodies of the recipient will react to the donor's red blood cells.

Blood group	Agglutinins (antigens) on surface of red blood cells	Antibodies in plasma
A	type A	anti B
B	type B	anti A
AB	type A and type B	none
O	none	anti A and anti B

Matching donors and recipients

- People of group O can donate blood to anyone as their red blood cells do not have any antigens, so the recipient's antibodies have nothing to react with.
- People of group AB can receive any type of blood as they do not have any antibodies in their plasma to react to donors' antigens.
- People of group A cannot receive group B blood because their anti B antibodies would coagulate it.
- People of group B cannot receive group A blood because their anti A antibodies would coagulate it.

People are also classified according to the rhesus factor. Your blood is rhesus positive if your plasma has a D protein, and rhesus negative if it does not. Rhesus-negative people cannot receive rhesus-positive blood as they would make antibodies against the D protein.

Questions

1 Who might need a blood transfusion? ↓E

2 How can the risk of blood clots be reduced?

3 Describe how blood clots when you cut yourself. ↓C

4 Why do you think people of blood group O are called universal donors?

5 Why do you think people of blood group AB are called universal recipients? ↓A*

6 Explain why people of blood group A cannot receive blood from donors of blood group B.

A Why did the first living organisms on Earth, early bacteria, respire anaerobically?

B How did photosynthesising bacteria alter the Earth's atmosphere?

▲ The leopard frog *Rana pipiens* lives in grasslands and woodlands throughout North America. It eats insects and sometimes small fish. It returns to water to breed.

Aerobic respiration

The first life forms on Earth, three and a half billion years ago, were ancient types of bacteria. These obtained energy from chemical reactions. Some used anaerobic respiration. There was no free oxygen in the Earth's atmosphere at that time. Then some bacteria developed the ability to photosynthesise. This released free oxygen into the atmosphere. Oxygen killed many of the anaerobic bacteria around at the time, but some survived. Some coped with oxygen, and most of the life forms that have since evolved use aerobic respiration. Some organisms can use both anaerobic and aerobic respiration, depending upon the conditions. You probably know that your muscle cells can respire anaerobically for a while.

Gaseous exchange

If an organism respires aerobically, it has to get oxygen to its cells (and then to the mitochondria in its cells). The waste carbon dioxide from aerobic respiration has to be removed from the organism as it is toxic; it would lower the pH and disrupt enzyme activity. The exchange of these two gases, into and out of the organism, is called gaseous exchange.

Single-celled organisms

Simple one-celled organisms such as aerobic bacteria and amoebae have a large surface area compared with their volume. There is enough cell surface membrane to allow sufficient oxygen to diffuse into the cell. The carbon dioxide produced can all diffuse out.

Earthworms

Earthworms have many cells, but they are long and thin. They have a large surface area compared to their volume. Oxygen diffuses through their thin, **permeable** skin and into their blood vessels. The blood carries oxygen from the skin to respiring cells. It also carries carbon dioxide from respiring cells to the skin. Earthworms, like all organisms, contain a lot of water. They do not have waterproof skin, as humans do, so they secrete mucus to stop themselves drying out. They also live in damp places.

Amphibians

Most of the gaseous exchange in a frog happens across its skin. Frogs also have simple lungs, and can obtain oxygen from the floor of their mouth.

Water loss

Because their skin is permeable to gases, frogs are susceptible to excessive water loss. To avoid drying out frogs need to live in damp places. Some survive in drier habitats by having a layer of slime over the skin.

Fish

Fish also have gills for gaseous exchange. Remember that fish have a single circulatory system.

- The blood flows from the heart to the gills.
- The gills have filaments which give a very large surface area.
- Each filament is well supplied with blood.

- As the fish swims, it gulps water into its mouth.
- Then with the mouth closed, it raises the floor of its mouth and forces the water out over the gills.
- Oxygen dissolved in the water diffuses into the fish's blood at the gills.
- Carbon dioxide diffuses from the blood in the gills into the water.
- The oxygenated blood then flows to the fish's body organs.

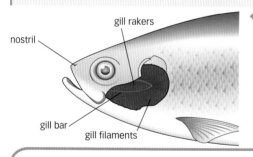

◀ The position of the gills in a fish. The gill flap is cut away. The gill bar supports the filaments.

Questions

1 Explain why all living organisms that respire aerobically need to have gaseous exchange.

2 Explain why larger complex organisms need a special surface for gaseous exchange.

3 Make a table to compare gaseous exchange in:
  (a) an earthworm (b) an amoeba
  (c) a frog (d) a fish

4 Describe how fish force water over their gills.

5 Explain why frogs are at risk of drying out.

E
C
A*

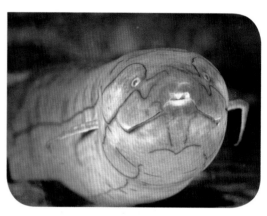

▲ The South American lung fish. It has a pair of lungs, one on either side of its throat.

Exam tip OCR

- Do not confuse gaseous exchange with breathing. Breathing is the ventilation movement that brings air to the gaseous exchange surface. Fish are ventilating when they open their mouths, gulp water, then close their mouths and open the gill flaps to force water over the gills. Gaseous exchange is the diffusion of oxygen and carbon dioxide into and out of the blood at the gills or lungs.

- Do not confuse either breathing or gaseous exchange with respiration. Respiration happens in the cells.

The main parts of the human respiratory system

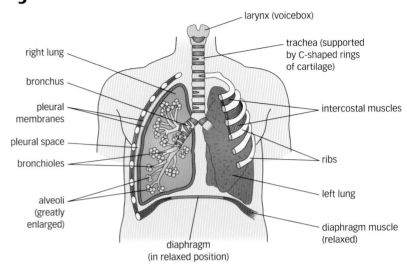

▲ This diagram shows a surface view of the left lung, and a section through the right lung showing the airways and air sacs inside. Pleural membranes cover the lung and the inside of the rib cage. The pleural space between them contains pleural fluid, which allows easy slippage of the moving lungs during breathing, and also helps prevent the lungs from collapsing.

Breathing (ventilation)

Your lungs are your gaseous exchange surface. The many millions of air sacs (alveoli) in the lungs give a large surface area for oxygen to diffuse into your blood and for carbon dioxide to leave your blood and be breathed out. **Breathing**, or **ventilation**, is how you get the air in and out of your lungs.

Breathing in (also called inhalation or inspiration)	Breathing out (also called exhalation or expiration)
• Your intercostal muscles contract and raise your rib cage up and out. • Your diaphragm flattens. • These two things increase the volume inside your chest and lungs. • The air pressure in your lungs is lower than outside. • So air enters the lungs from outside. It passes along the trachea, bronchi, and bronchioles to the alveoli.	• Your intercostal muscles relax and your rib cage lowers. • Your diaphragm domes upwards. • These two things reduce the volume in your chest. • The air pressure in your lungs is greater than outside. Your elastic alveoli also recoil (snap back) to normal size. • So air is pushed out from your lungs to outside. However, some air stays – this is **residual air**. If it didn't stay, your alveoli would close up.

Using a spirometer

Doctors may measure a patient's lung function, using a spirometer. The patient breathes while attached to a machine. It can measure

- **tidal volume** – the volume of air you breathe in, in one breath
- **vital capacity** – the maximum volume of air you can breathe out after taking a big breath in
- **lung capacity** – your vital capacity plus your residual air.

The alveoli

The alveoli form your gaseous exchange surface. They link your blood to the air. At the alveoli:

- Oxygen diffuses from the alveoli into the bloodstream.
- Carbon dioxide that has entered your blood at your body tissues diffuses from your blood into the alveoli, to be breathed out.

Like all of your organs and tissues, the alveoli and the lungs are moist. Your cells are about 70% water, as all the chemical reactions inside them take place in solution. Oxygen actually diffuses quicker when not in solution, but your lungs are moist. Fortunately, oxygen will dissolve in the very thin film of moisture and will still diffuse in solution.

The alveoli are surrounded by blood capillaries. The walls of your alveoli are just one single layer of cells. The walls of your capillaries are also just one single layer of cells. So the oxygen and carbon dioxide do not have far to diffuse. Alveoli are adapted for efficient gaseous exchange because they

- are permeable
- have thin walls
- have a large surface area
- have a good blood supply.

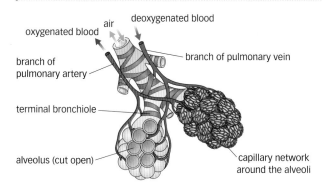

▲ How the alveoli and capillaries in the lungs aid gaseous exchange

A Describe inspiration (how you breathe in).

B Describe expiration (how you breathe out).

Questions

1 Where does gaseous exchange happen in humans?

2 When you breathe out, the exhaled air contains much more carbon dioxide than inhaled air. Explain where the extra carbon dioxide has come from and how it got to the lungs.

3 Explain how your gaseous exchange surface is well adapted for efficient gaseous exchange.

4 Look at the diagram below. It is a spirometer trace on graph paper. For the person whose trace this is, find:
(a) their tidal volume
(b) their vital capacity.

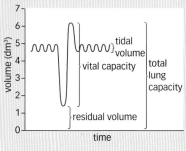

▲ A spirometer trace for a 15-year-old boy

5 Explain why your vital capacity is not the same as your lung capacity.

Learning objectives

After studying this topic, you should be able to:

- ✔ know about diseases of the respiratory system, including pneumonia, asthma, and lung cancer, and their causes
- ✔ explain what happens during an asthma attack

Key words

constrict

Did you know...?

People often say that cancerous cells divide very quickly. This is not true. They do not divide any quicker than normal dividing cells; they just do not 'know' when to stop dividing. This is because something, such as the tar in tobacco smoke, has caused a mutation to the genes that control cell division. Lung cancer is slow growing. It takes up to 30 years for a lung cancer tumour to become large enough to detect. Smokers may think they are fine after 20 years of smoking, but the tumour may be growing slowly and not yet causing symptoms.

A How is the respiratory system protected from infection?

B Explain why damaged cilia can make you more prone to lung infection.

How the respiratory system protects itself

You have hairs in your nose which can trap large particles of dirt in the air you breathe in. However, small particles and pathogens may get past these hairs.

There is a layer of cells lining the trachea and bronchi. Some of these cells have cilia and some secrete mucus.
- The mucus traps small particles and pathogens such as bacteria, viruses, and fungal spores.
- The cilia beat and waft the mucus up to the back of the throat.
- Once there it can be swallowed, and the stomach acid kills the trapped pathogens. Or it can be removed by coughing or blowing your nose.

You also have special white blood cells, called **macrophages**, that squeeze out of capillaries and patrol the lung tissues. They ingest foreign particles and some pathogens.

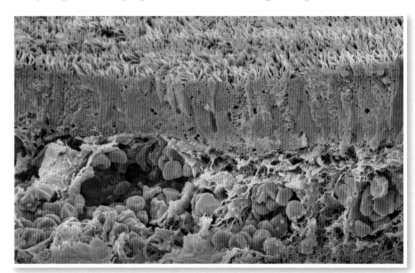

▲ Section through the lining of the trachea, seen with a scanning electron microscope (×600). False colour has been added.

What happens if this sweeping system is not working?

The airways end in the lungs and go no further – they are a dead end. Any pathogens entering the lungs will cause infection unless they are trapped and wafted back up the airways, out of the lungs. If the cilia are not working properly, small particles and pathogens remain in the lungs and can cause disease.

Some respiratory diseases

The table lists common respiratory diseases. Some of these are digested by specific enzymes, to smaller soluble molecules, in certain parts of the digestive system.

Disease	Cause	Symptoms
Bronchitis	Virus or bacteria, can also be triggered by breathing in smoke	Cough (often bringing up a yellow-grey mucus), sore throat, wheezing, blocked nose.
Asbestosis	Asbestos fibres trapped in the alveoli. This is an occupational disease. Some people have been exposed to the fibres during their work.	Inflammation and scarring of the alveoli, leading to difficulty breathing and reduced gaseous exchange. May lead to cancer.
Cystic fibrosis	Genetic and inherited	Cells lining the airways are affected. Thick mucus is secreted. Cilia are not hydrated enough and cannot waft. Mucus with trapped pathogens builds up, and this leads to chest infections and reduces gaseous exchange. Lungs eventually become damaged.
Lung cancer	Most commonly tar in tobacco smoke	Cells lining the bronchioles keep on dividing, forming a tumour. This reduces the surface area for gaseous exchange, and causes chest pain and a prolonged cough, with blood.
Asthma	Inhaling pollen or other allergens, infection, cold air, hard exercise, or stress	Difficulty breathing, wheezing, tight chest. Can be treated with bronchodilator drugs taken via an inhaler. Lining of the airways becomes inflamed, causing a build up of fluid. The muscles in the bronchi and bronchioles contract. This makes the bronchi and bronchioles constrict (become narrower), restricting the airways.

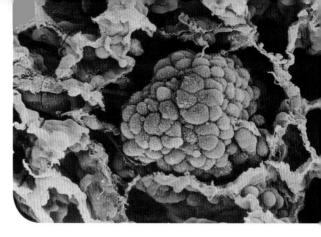

▲ Lung cancer. Coloured scanning electron micrograph of a small cancerous tumour filling an alveolus of the lung (× 400). Some of the cancer cells have separated from the main tumour. If they enter the blood, they may be carried to other tissues and set up secondary tumours.

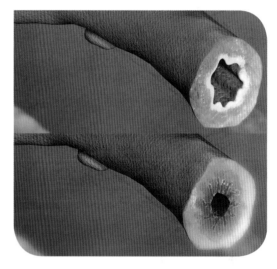

▲ A section through a normal bronchus (top) and an inflamed, **constricted** bronchus during an asthmatic attack (bottom)

Questions

1 Describe what happens during an asthma attack.
2 Describe how lung cancer may develop.
3 Explain how some people developed asbestosis after working for a long time where they were exposed to asbestos fibres.

C

Learning objectives

After studying this topic, you should be able to:

✔ explain why food has to be digested

✔ explain the role of enzymes in chemical digestion

✔ describe the functions of the parts of the human digestive system

✔ explain how digested food molecules are absorbed into the blood

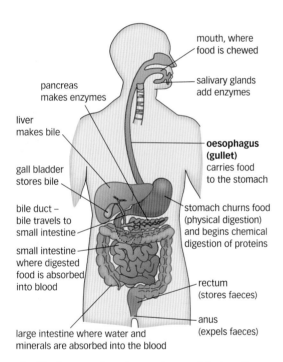

mouth, where food is chewed

salivary glands add enzymes

pancreas makes enzymes

liver makes bile

oesophagus (gullet) carries food to the stomach

gall bladder stores bile

bile duct – bile travels to small intestine

stomach churns food (physical digestion) and begins chemical digestion of proteins

small intestine where digested food is absorbed into blood

rectum (stores faeces)

anus (expels faeces)

large intestine where water and minerals are absorbed into the blood

▲ The human digestive system and its functions

Key words

mouth, stomach, carbohydrase, lipase, protease, small intestine, large intestine

What happens to the food you eat?

While food is in your gut, it is still really outside your body. The large molecules in the food have to be broken down (digested) into smaller molecules so that they can pass across your gut wall and into your blood or lymph. Then your circulatory system takes the digested food to your cells and tissues. Here, you use the food for energy or growth and repair.

Physical digestion

Chewing food in your **mouth** and squeezing food in your **stomach** are both forms of physical digestion. The resulting smaller pieces of food can move more easily through the rest of the digestive system.

Chemical digestion

Carbohydrates, fats, and proteins are digested by specific enzymes in certain parts of the digestive system.

Food	Type of enzyme	Part of gut where enzyme works	Products of digestion
carbohydrates	**carbohydrases**	mouth and small intestine	starch, converted to maltose and then to glucose, a simple sugar
fats (lipids)	**lipases**	small intestine	fatty acids and glycerol
proteins	**proteases**	stomach and small intestine	amino acids

Enzymes have a specific optimum pH

You have hydrochloric acid in your stomach, giving it a very low pH of between 1 and 2. This is primarily to kill any pathogens in your food. The protease enzyme in your stomach is well adapted to this low pH and will not work at higher pH values. However, other enzymes in your mouth and small intestine work best at higher pH values of between 7 and 8.

Bile

To help you digest fats in the small intestine, your liver makes bile. Bile is stored in your gall bladder and released into the small intestine. Bile emulsifies the fats (breaks the fats into smaller droplets). This gives the lipase enzymes more surface area to work on.

A Describe the functions of each of the following:
(a) oesophagus (b) stomach (c) small intestine (d) large intestine.

B Explain the difference between physical and chemical digestion.

Absorption of digested food in the small intestine

The products of digestion diffuse across the wall of the **small intestine** into the blood plasma or lymph.

Adaptations of the small intestine

The small intestine is well adapted to absorb digested food efficiently:

- It is very long.
- It has a large surface area because its lining is folded and has finger-like projections called villi.
- The cells covering each villus have microvilli, which increase the surface area even more.
- The lining is thin.
- There is a good blood supply.

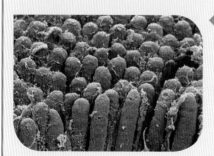

◀ Coloured scanning electron micrograph of villi in the small intestine (× 100)

The large intestine

The **large intestine** absorbs water and some minerals into the blood. The semi-solid waste (faeces) that is left in the large intestine is then passed out of the anus. This is called egestion.

Questions

1 What is the function of the large intestine? E

2 Where in your digestive system are each of the following chemically digested? (a) carbohydrates (b) fats (c) proteins

3 Digested food enters your blood in your small intestine and leaves your blood at body tissues. Why does it leave your blood at your body tissues? C

4 What is the function of bile in digestion?

5 The pH in your stomach is around 1.2 and in your small intestine it is about 7.8. Proteins are digested in both places. Do you think the same protease enzyme works in both places? Explain your answer. A*

6 Explain how the small intestine is well adapted to efficiently absorb digested food.

11: Excretion

Learning objectives

After studying this topic, you should be able to:

✔ understand the difference between egestion and excretion

✔ name and locate the main organs of excretion and name the substances they excrete

✔ explain why substances have to be excreted

Key words

urea

A Explain the difference between egestion and excretion.

B Name four organs of excretion.

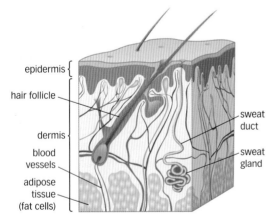

epidermis
hair follicle
dermis
blood vessels
adipose tissue (fat cells)
sweat duct
sweat gland

▲ A section of human skin

Did you know...?

The skin is your largest organ. It weighs about 4–5 kg.

Waste disposal

During digestion, the semi-solid waste (faeces) that is left leaves your body via your anus. This is egestion, not excretion – this waste was not made in your body.

Some chemical reactions in your cells make toxic waste products. Respiration produces water and carbon dioxide. Your body has to remove toxins, otherwise you would be poisoned. Your body also has to regulate the amount of salts and water in it. It is important that the concentration of water molecules in your blood plasma is kept constant. Too much water, and blood cells would swell and burst. Too little water, or too much salt, and they would shrivel and not function. Your nerves would not work properly if there were too many or too few salts in your body.

The main organs of excretion

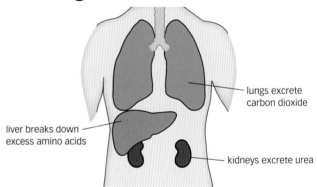

lungs excrete carbon dioxide

liver breaks down excess amino acids

kidneys excrete urea

▲ The main organs of excretion and their functions

The skin

Your skin makes sweat to cool you. The water in the sweat uses your body heat to evaporate. Sweating also gets rid of excess water and salts.

The lungs

Your respiring cells make carbon dioxide. It is carried in the blood to the lungs and then diffuses into the alveoli to be breathed out. So your lungs excrete carbon dioxide that was made in your cells. The carbon dioxide would otherwise poison you because it would lower your blood pH. Your enzymes would not work properly and you would die.

As carbon dioxide levels in your blood increase they are detected by the brain. The brain then increases your breathing rate to remove the carbon dioxide more quickly.

The liver

Your liver breaks down old red blood cells. The chemicals from them go into the bile and pass out with the faeces. Your liver also breaks down hormones and medicines or other drugs, such as alcohol.

If you have eaten more protein than your body needs, the liver breaks down excess amino acids into ammonia. Ammonia has a high pH and is very soluble, so if it got into your blood it would be very harmful. Enzymes would not be able to function. In your liver cells the ammonia reacts with carbon dioxide (another waste product) to make **urea**. Urea is toxic, but not as toxic as ammonia. Your blood carries the urea to your kidneys, which remove it.

The kidneys

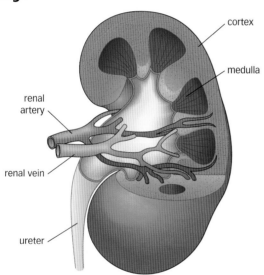

▲ The structure of the kidney. Blood flows in through the renal artery, is filtered and leaves the kidney in the renal vein. The waste urine, containing water, salts, and urea, passes down the ureter to the bladder.

Each of your kidneys has about one million filtering units. Blood enters your kidney in the renal artery. It is filtered under high pressure and lots of substances are filtered out:
- glucose
- salts
- water
- urea.

Then the useful substances:
- all the glucose
- some salts
- some water

are reabsorbed into the blood. The remaining liquid, called urine, passes down the ureter to the bladder. It is stored in the bladder and passed out when convenient.

As well as removing urea, your kidney also regulates the amount of salts and water in your body.

Exam tip

- ✔ Remember that it is the water in sweat that evaporates, not the sweat itself. The evaporation changes water from a liquid to a gas, and that takes heat energy from your skin, blood, and body. This is how sweating cools you.

Questions

1 For each organ you named in Question B, state its main excretory product. ↓ E

2 Explain why carbon dioxide has to be removed from the body.

3 How do you think the amount of urea in your urine would change if you started eating a high protein diet? Explain your answer.

4 How do you think the water content of your urine would change if you drank a lot of tea and lemon squash? Explain your answer. ↓ C

5 How do you think the water content of your urine would change on a hot day if you ran around and did not drink any extra water? Explain your answer.

6 Explain how your breathing rate is made to increase after exercise. ↓ A*

Learning objectives

After studying this topic, you should be able to:

✔ explain how the structure of the kidney tubule enables it to filter blood and produce urine

✔ explain how the amount of water in the blood is regulated

✔ explain the principle of the dialysis machine

Key words

glomerulus, **selective reabsorption**, **afferent arteriole**, **efferent arteriole**, **pituitary gland**

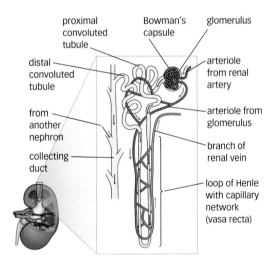

▲ The position of nephrons in the kidney, and the structure of a nephron

A Why is it important that salts and water in the blood are carefully regulated?

The nephron

You have learnt about the gross structure of a kidney. There are about a million filtering units, called nephrons, in each kidney. Each nephron consists of

- a knot of capillaries, called a **glomerulus**, inside a capsule, where high pressure filtration occurs
- a region for **selective reabsorption**, where useful substances eg glucose pass into the blood
- a region for salt and water regulation.

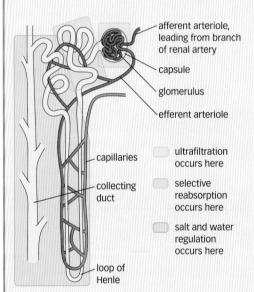

◀ Detailed structure of a nephron showing the regions for high pressure filtration, selective reabsorption, and salt and water regulation. The tubule is surrounded by capillaries so that substances can easily pass back into the blood.

High pressure filtration

Blood enters your kidney, in the renal artery, under high pressure. Many arterioles branch off from the renal artery and one arteriole goes to each glomerulus. Blood goes from the **afferent arteriole** into the glomerulus. It then leaves the glomerulus in another arteriole, called the **efferent arteriole**.

The efferent arteriole has a narrower diameter than the afferent arteriole. This produces a bottleneck effect. The blood cannot leave the glomerulus as fast as it is entering, so it is under high pressure, and the capillaries of the glomerulus are very leaky. The result is high pressure filtration. Substances with small molecules – water, salts, urea, glucose, amino acids, vitamins, and spent hormones – are filtered out of the blood. They pass along the tubules of the nephron dissolved in liquid that was squeezed out from the glomerulus.

From the first part of the nephron, all the useful substances – glucose, amino acids, vitamins, some salts, and some water – are reabsorbed by selective reabsorption. The loop of Henle, the rest of the tubule, and the collecting duct regulate the amount of salt and water in the body.

- If your blood is very watery, less water is reabsorbed from the kidney tubules and a lot of dilute urine is produced.
- If your blood is not very watery, more water is reabsorbed from the kidney tubules and a smaller volume of concentrated urine is produced.

Antidiuretic hormone (ADH)

The hormone ADH is released from a gland in the brain, your **pituitary gland**, directly into the blood. Your blood carries it to its target organs, the kidneys. This hormone makes the walls of the collecting duct more permeable to water, so more water can be reabsorbed into the blood.

A negative feedback mechanism is involved. As your blood passes through your brain, your hypothalamus detects how watery it is.

- If your blood is watery, less ADH is released. Less water is reabsorbed in the kidneys and more water is lost in urine. This adjusts the water content of your blood.
- If your blood is not very watery, more ADH is released. More water is reabsorbed in the kidneys and less water is lost in urine.

Renal dialysis

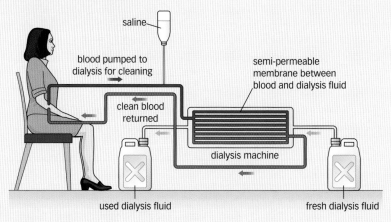

saline

blood pumped to dialysis for cleaning

clean blood returned

semi-permeable membrane between blood and dialysis fluid

dialysis machine

used dialysis fluid

fresh dialysis fluid

▲ Sometimes people's kidneys stop working properly. When this happens their blood can be filtered by a dialysis machine.

Exam tip OCR

- ✔ How can you remember when more ADH is released? Diuresis means *making urine*. Anti- means *against*. Antidiuretic hormone reduces the volume of urine. More ADH is released if your blood/body needs to conserve water – if you have been sweating and/or not drinking much. Less ADH is released if your blood is watery – if you have been drinking a lot.
- ✔ How can you remember which is the afferent and which is the efferent arteriole? *Affere* is Latin and means to carry *towards*; *effere* is Latin and means to carry *away from*.

Questions

1 What substances are reabsorbed into the blood during selective reabsorption?

2 Explain how high pressure filtration occurs in the glomerulus.

3 Describe how ADH release is controlled using negative feedback.

A*

Growth and development

At birth we can tell what sex the child is because of the external **genitals** – the primary sexual characteristics. At puberty, children's bodies begin to change into those of sexually mature adults. This stage of a person's life is called adolescence, and lasts several years. The first thing that happens is that the ovaries in females, and testes in males, develop and begin to produce the **sex hormones**.

The female sex hormones oestrogen and progesterone are made in the ovaries. They are involved in controlling the menstrual cycle.

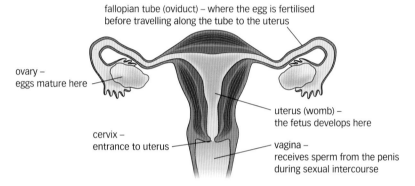

fallopian tube (oviduct) – where the egg is fertilised before travelling along the tube to the uterus

ovary – eggs mature here

uterus (womb) – the fetus develops here

cervix – entrance to uterus

vagina – receives sperm from the penis during sexual intercourse

▲ The female reproductive system and its functions

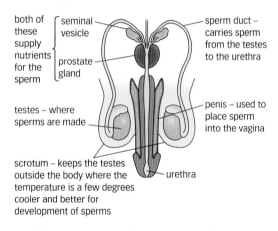

both of these supply nutrients for the sperm { seminal vesicle, prostate gland }

sperm duct – carries sperm from the testes to the urethra

testes – where sperms are made

penis – used to place sperm into the vagina

scrotum – keeps the testes outside the body where the temperature is a few degrees cooler and better for development of sperms

urethra

▲ The male reproductive system and its functions

Secondary sexual characteristics

At puberty, your sex hormones cause your secondary sexual characteristics to develop.

Male secondary sexual characteristics	Female secondary sexual characteristics
• The voice breaks (deepens). • Hair grows on the face and body. • The body becomes more muscular. • Genitals develop. • The testes start making sperm.	• The breasts develop. • Pubic hair and hair under the arms grows. • The hips widen. • Periods start (menstruation) and eggs mature.

A Name the female and male sex hormones.

B List four secondary sexual characteristics for males and four for females.

The menstrual cycle

At puberty, females begin to have a menstrual period each month. They have a monthly menstrual cycle.

Several hormones help coordinate the menstrual cycle.
- The pituitary gland in the brain releases a hormone called FSH (follicle stimulating hormone).
- FSH causes an egg in one of the ovaries to mature.
- It also stimulates the ovaries to make the hormone oestrogen.
- Oestrogen stimulates the pituitary gland to release another hormone, LH (luteinising hormone).
- LH triggers the release of the egg (**ovulation**) from the ovary.
- Oestrogen also inhibits further production of FSH and it repairs the uterus lining (the endometrium, which is the innermost layer of the uterus wall).
- Progesterone maintains the uterus lining and inhibits LH.

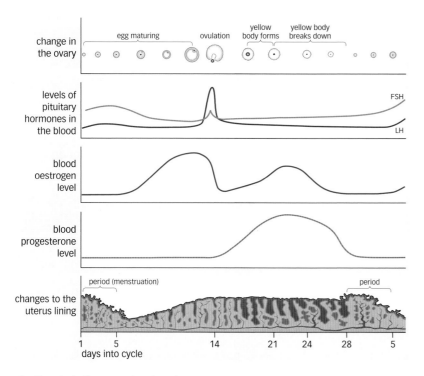

▲ Events in the menstrual cycle

Conception happens if a sperm meets the egg and fertilises it. If the egg is not fertilised, at the end of the cycle the uterus lining passes out of the body. This is the period. If the egg is fertilised then the uterus lining stays so that the fetus can develop.

Key words

genitals, sex hormones, ovulation, conception

Exam tip

✓ It may seem complicated, but you need to learn the menstrual cycle and the names and functions of all the hormones involved.

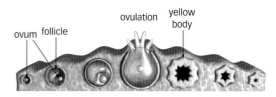

▲ Changes in the ovary during the menstrual cycle. The egg develops in a follicle. It then bursts out of the follicle. The empty follicle develops into a yellow body which makes progesterone; this stops menstruation.

Questions

1 Where is FSH?
2 What does FSH do?
3 What is the function of progesterone?
4 What is the role of oestrogen in the menstrual cycle?
5 During which part of the menstrual cycle is a woman most likely to conceive?
6 When is the level of LH highest?
7 Explain why high levels of the hormone progesterone are needed during pregnancy.

▲ Contraceptive pills in a blister pack. Each pack contains enough pills for one month. They are usually taken for 21 days of each month and then not taken for 7 days, so the woman has a period.

Key words

contraception, IVF, amniocentesis, miscarriage

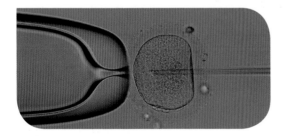

▲ A human sperm being injected into a human egg

Controlling fertility with female hormones

Humans can control fertility by using female sex hormones. These hormones can be used to reduce fertility or to promote fertility.

Reducing fertility

During pregnancy, both oestrogen and progesterone levels are high and they inhibit FSH and LH production from the pituitary gland. This prevents the development and release of any more eggs.

Scientists realised that if women took oestrogen and progesterone in a daily pill, the high levels in the body would mimic pregnancy and prevent ovulation. Without ovulation you cannot become pregnant. Preventing pregnancy is called **contraception**.

The first contraceptive (birth-control) pills contained high amounts of oestrogen. They prevented ovulation but many women suffered from side-effects. Contraceptive pills now contain a much lower dose of oestrogen and some progesterone, or just progesterone. These give fewer side-effects.

Increasing fertility

Some couples are infertile and therefore cannot conceive. Possible causes of infertility are:

- blocked fallopian tubes or sperm ducts
- eggs do not develop or are not released from ovaries
- the testes do not produce enough sperms.

Fertility treatment may help.

A procedure called in vitro fertilisation (**IVF**) may be used to treat women who cannot become pregnant naturally. *In vitro* means 'in glass'. In IVF:

- The woman is injected with FSH, which stimulates her ovaries to produce eggs.
- The eggs are then collected from the woman and mixed with the man's sperm in a glass dish.
- To make the procedure more likely to work, healthy sperms are selected and one is injected into each egg.
- The fertilised eggs begin to develop into embryos.
- When they are tiny balls of cells, two are chosen and inserted into the woman's uterus.

Other treatments for infertility

Treatment	Description and reason for treatment
Artificial insemination	If the male's sperm count is low or the woman's vagina is hostile to her partner's sperm, sperm can be inserted into the woman's vagina. Also used for single women or lesbian couples who wish to become pregnant.
Egg donation	A woman may donate some of her eggs to another woman who cannot make eggs. Women undergoing IVF often donate 'spare eggs'.
Surrogacy	If a woman has had her uterus removed, another woman (the surrogate mother) may have the first woman's embryo (the result of IVF) implanted in her uterus. After the birth the surrogate mother gives the baby back to its biological mother.
Ovary transplants	So far only a few ovary transplants have been carried out, with the donor being the identical twin of the recipient. This could restore the fertility of women who undergo early menopause and are no longer fertile, or who have had radiation treatment for cancer. A woman could have her ovaries removed before being treated for cancer, have them frozen and have them transplanted back after the treatment.

Being childless can cause distress and sadness to people who want a family, and fertility treatments allow some people to have a child who could not do so otherwise. However, these treatments are very expensive and there is no guarantee they will work. All medical procedures carry some risk.

Checking fetal development

Ultrasound scans are used to check fetal development. However, some abnormalities cannot be seen on a scan. Down's syndrome is caused by the presence of an extra chromosome. To check for Down's syndrome, doctors can insert a needle into the uterus and take some amniotic fluid containing fetal cells. This procedure is called **amniocentesis**. The cells are grown in a lab so that they divide. Their chromosomes can be observed under a microscope and counted.

If the fetus has an extra chromosome, the couple have to decide whether to terminate the pregnancy. Some people think that screening out disabilities means that disabled people are undervalued in society. There is a 1% (1 in 100) chance that amniocentesis could cause a **miscarriage** and could abort a healthy fetus.

A What does IVF stand for?

B Explain how FSH can be used to increase fertility in women who cannot conceive (become pregnant).

Exam tip OCR

✔ If you are using abbreviations like IVF, give the full name first with the abbreviation after it in brackets. Then use the abbreviation.

✔ If you are asked to discuss ethical issues, ensure that you give a balanced argument and don't dwell on your personal views.

Questions

1 Name two female hormones produced in the ovaries.

2 Discuss the advantages and disadvantages of: (a) IVF (b) egg donation (c) surrogacy (d) ovary transplants.

3 Some fertility drugs contain a chemical that inhibits oestrogen. How do you think this might increase fertility?

4 Some families have a history of a particular genetic disorder. In these cases, fertilisation can be in vitro and the embryos can be tested for genetic defects. Only healthy embryos will be implanted. Do you think this is a good idea? Give reasons for your answer.

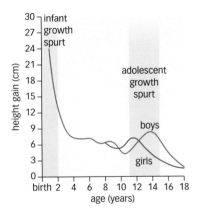

▲ The growth rate, as height gain per year, from birth to 18 years

Charts like the one above are based on an average of the population. Not everyone's growth will follow this pattern.

Monitoring growth and development

Before you were born, doctors and midwives used ultrasound scans to monitor your development in the uterus. They monitored your rate of growth and increase in head size as well as your heart rate. This told them if you were growing and developing at a normal rate.

At birth and during the first six months, the midwife measures a baby's

- head circumference
- body length
- mass

and plots them on average growth charts. The baby's growth is compared with the range of average values. This regular monitoring can alert the midwife if you have any growth problems.

If a baby is growing too slowly, its pituitary gland may not be producing enough growth hormone. Growth hormone stimulates general growth, especially that of long bones and muscles. A deficiency can be treated with injections of human growth hormone.

Stages of human growth

We do not grow at a constant rate. The graph on the left shows the stages of human growth.

infancy	– First two years of life – Highest rate of growth, gaining around 15–24 cm in a year
childhood	– From 2–11 years of age, until puberty starts – Growth occurs at a slower rate than during infancy
adolescence	– From 11–15 years of age, when puberty begins – Growth spurt for girls aged 10–12 and boys aged 12–15 years
maturity	– Males may continue to grow until the age of 18–20 years – Most females reach their full adult height by 16 years of age
old age	– Above 60–65 years – Physical abilities start to deteriorate

> **A** Name and describe the main stages of the human growth curve.

Key words

infancy, childhood, adolescence, maturity, life expectancy, old age

Your final height is determined by

- your genes – many genes determine your height potential
- your diet – you need good quality protein for growth and enough food for your energy needs
- the amount of exercise you do – exercise stimulates growth of bones and muscles
- your hormones – growth hormone, thyroxine, and insulin
- health and disease – if you are often ill or do not get enough sleep you may not grow properly.

Life expectancy

In the developed world, **life expectancy** has increased. We expect to live to between 75 and 80 years. This does not mean everyone will live this long. Many die earlier and some later. **Old age** is officially above 60–65 years but many older people carry on working and living full and active lives.

Life expectancy has increased in developed countries because we have

- less industrial disease such as asbestosis, and fewer accidents in the workplace
- healthier diets, with few cases of vitamin or mineral deficiencies
- better housing
- improved lifestyle so people can have a positive outlook on life
- vaccinations to prevent many infectious diseases
- better treatments for cancer and heart disease.

However, there may be problems with increased life expectancy:

- a large ageing population – there are more people aged over 65 than under 16 in the UK
- many of these will need medical treatment or care
- this could affect the job prospects for younger people as more older people need to keep working for economic reasons and/or because they want to
- the state pension system will need to be redesigned as it cannot afford to pay pensions for 30 or more years to people; the retirement age will also be raised.

▲ Increased life expectancy has led to a large ageing population

Questions

1 What factors determine your final height? ↓ E

2 Use the graph of growth rate on the previous page to answer the following questions. What is the average height gain per year: (a) aged 1 year (b) aged 2 years (c) aged 6 years (d) for boys aged 14 years? ↓ C

3 Why do you think more people are living longer in the UK today, compared with 50–100 years ago? What possible problems may arise as a result? ↓ A*

Key words

donor, tissue match, immunosuppressant drugs

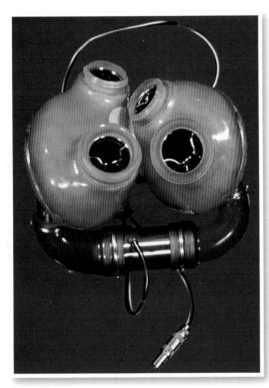

▲ The Jarvik-7 artificial heart was developed by Dr Robert Jarvik and a biomedical engineer, Dr Lyman. It is made from polyurethane and titanium. The inside is smooth and seamless so it does not cause blood clots that would lead to strokes. Many people have had this type of artificial heart while waiting for a heart transplant. However, the wires protrude through the skin.

Mechanical replacements

Many mechanical replacements for body parts are used outside the body, for example

- heart–lung machines, used during open heart surgery to divert blood from the heart
- kidney dialysis machines, used to filter the blood of people with renal (kidney) failure, removing urea, excess salts, and water
- mechanical ventilators, used to aid breathing in patients whose rib cage muscles are paralysed.

▲ A renal dialysis machine in action

Other mechanical devices can be implanted into the body. These include:

- heart pacemakers
- artificial hearts
- artificial knee and hip joints
- eye lenses.

When medical engineers are designing these implants they need to consider certain factors, such as:

- size – this is why there is no artificial kidney that can be implanted; dialysis machines are very large
- battery life (if powered) – pacemaker batteries last 7–10 years
- body reactions – inert materials that do not react with body fluids are used to construct implants; artificial hearts are made of titanium and plastic
- strength – titanium is used for artificial joints.

> **A** Describe two mechanical organ replacements that can be used (a) outside, and (b) inside the body. Explain why they are used.

Organ transplants

Many body parts can be replaced by transplanting donated organs.

Blood transfusions	Successfully carried out for over 100 years, using blood from live **donors** that is stored in blood banks. Blood types must be matched.
Cornea transplants	Also known as corneal grafts. A cornea removed from a recently dead donor is transplanted into the recipient's eye. No risk of rejection as the cornea has no blood vessels.
Heart transplants	First carried out in 1967 in South Africa. Donors are usually recently dead, but may be living – if someone has a heart–lung transplant to replace diseased lungs, the recipient's healthy heart can be donated to someone else.
Lung transplants	Lungs from a recently dead donor may be transplanted into a recipient suffering from cystic fibrosis, for example. Usually a heart–lung transplant is carried out.
Kidney transplants	Donor may be dead or living, as we can survive with only one kidney. A close relative may donate a kidney to a recipient.
Bone marrow transplants	Used to treat leukaemia. Living donors are tissue-typed and recorded on a register so that they can be matched to a recipient. They then give bone marrow.

All donors need to be a good tissue match and the right age and size for the recipients. Living donors have to be healthy and willing to donate. There is currently a shortage of organ donors.

Rejection

With most transplants there is a risk that the recipient's immune system will reject the transplant. **Tissue matching**, matching the donor and recipient's tissue type, and **immunosuppressant drugs** both reduce the risk of rejection. However, these drugs increase the risk of infections.

Did you know...?

The success rate for adult bone marrow transplants is about 4/10, and 7/10 in children. The recipient is receiving a donated immune system. Sometimes the transplanted marrow rejects the recipient – this is called graft versus host reaction and may be fatal.

Exam tip

- ✔ When discussing ethical issues, be objective and give some pros and some cons.

Questions

1 Explain why tissue types for donor and recipients have to be matched for most organ/tissue transplants.

2 Explain how a living heart donor may be used.

3 Discuss the ethical issues concerning organ/tissue transplants.

4 Why is there no problem of rejection with cornea grafts?

5 What problems are associated with taking immunosuppressant drugs?

Module summary

Revision checklist

- Some animals have no hard skeleton. Some have an external skeleton and some have an internal skeleton.
- Joints and muscles allow you to move. Muscles work in antagonistic pairs.
- Multicellular animals have a circulatory system to transport material to and from cells.
- Mammals have a double circulatory system consisting of heart and blood vessels (the cardiovascular system) and lymph.
- The cardiac cycle describes the events of each heartbeat.
- Artificial pacemakers, valve replacements, surgery, and heart transplants may treat disorders of the heart.
- Blood can be transfused. Blood groups have to be compatible.
- Organisms have special surfaces for gaseous exchange, such as skin, gills, and lungs, to obtain oxygen.
- Breathing moves air into and out of the lungs. Alveoli give a large surface area for gaseous exchange. Respiratory diseases include pneumonia, asthma, lung cancer, cystic fibrosis, and asbestosis.
- Food is digested to smaller molecules to be absorbed into the blood. Chewing is physical digestion; enzymes cause chemical digestion.
- Excretion is the removal of toxic waste made in the body. The skin excretes sweat; lungs excrete carbon dioxide; the kidney excretes urea, excess water, and excess salts. Urea is made in the liver from excess amino acids.
- Each kidney contains many filtering units called nephrons. Antidiuretic hormone controls how much water is reabsorbed. Renal dialysis can treat people with kidney failure.
- Secondary sexual characteristics and male and female reproductive organs develop at puberty. Hormones play an important part in growing up.
- Many treatments are available to treat infertility. Female hormones can also be used in contraceptives. Fetal development can be checked using ultrasound scans and amniocentesis.
- Humans pass through many stages of development: infancy, childhood, adolescence, and adulthood. Rate of growth depends on genes, diet, hormones, and health status.
- Some parts of the body can be replaced with mechanical devices or with organ or tissue transplants.

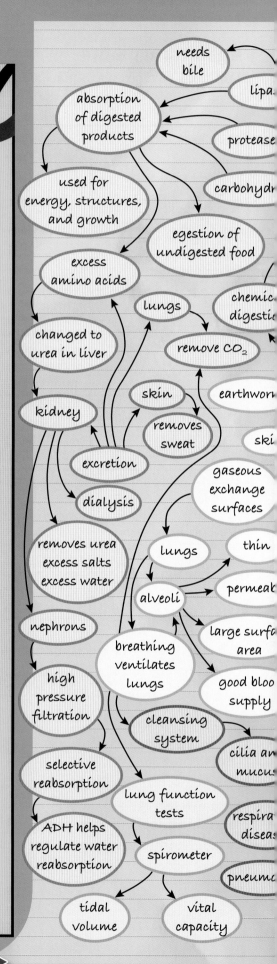

NOW USE THE B5 GRADE CHECKER ON PAGE 248

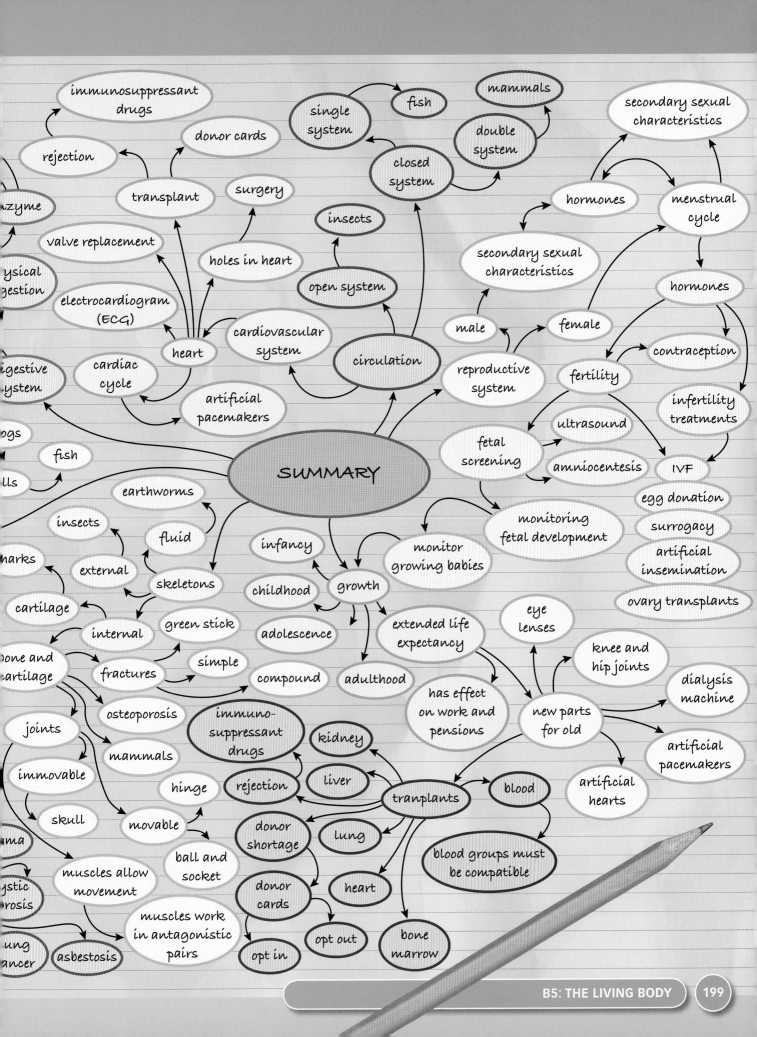

immunosuppressant drugs

donor cards

rejection

transplant

surgery

valve replacement

holes in heart

...zyme

electrocardiogram (ECG)

...ysical ...gestion

heart

cardiovascular system

...igestive ...ystem

cardiac cycle

artificial pacemakers

single system

fish

mammals

double system

closed system

open system

insects

circulation

secondary sexual characteristics

hormones

menstrual cycle

secondary sexual characteristics

hormones

male

female

contraception

reproductive system

fertility

infertility treatments

fetal screening

ultrasound

amniocentesis

IVF

egg donation

surrogacy

artificial insemination

ovary transplants

monitoring fetal development

SUMMARY

...ogs

fish

...lls

earthworms

insects

...harks

external

cartilage

internal

...one and ...artilage

joints

immovable

skull

...ama

...ystic ...rosis

...ung ...ancer

fluid

skeletons

green stick

fractures

simple

osteoporosis

mammals

hinge

movable

ball and socket

muscles allow movement

asbestosis

muscles work in antagonistic pairs

infancy

childhood

growth

adolescence

compound

adulthood

monitor growing babies

extended life expectancy

eye lenses

has effect on work and pensions

new parts for old

immuno-suppressant drugs

kidney

liver

rejection

tranplants

blood

donor shortage

lung

heart

donor cards

opt in

opt out

bone marrow

blood groups must be compatible

knee and hip joints

dialysis machine

artificial pacemakers

artificial hearts

OCR gateway Upgrade

Answering Extended Writing questions

Describe the structure of a long bone. Explain how bones can be broken and why elderly people are particularly prone to fractures. What are the risks when someone has a suspected fracture?

The quality of written communication will be assessed in your answer to this question.

Bones are hard because you need calcium for them. That's why you should drink lots of milk but I don't like it. Old people's bones break easily. You have to get the bone set in plaster.

↓ E

Examiner: This response has not answered the question. It has stated that bones are hard, but has not described their structure. The answer does not explain why older people's bones may break more easily; it just repeats the question. The risks associated with a suspected fracture have not been given. The reference to plaster, although true, is not relevant. The spelling, punctuation, and grammar are good.

There is marrow inside bones where red blood cells are made. At the ends there is cartilage. This is so joints work smoothly. Bone is living tissue. Bones break if you knock them sharply. Some people, specially old people have weak bones so they break more often. their bones don't have enough calcium. If you break your arm and the bone sticks out through the skin, it could get infected.

↓ C

Examiner: This answer shows some knowledge of bone structure. There is an explanation of why older people suffer more fractures, and a risk from a fracture is described. However, if the bone is sticking out through the skin as the candidate describes, this is a definite fracture and not a suspected one. There are a few grammatical errors.

Long bones have a shaft and inside is marrow where blood cells are made. At each end there is cartilage which is smooth and slippery. It helps joints work.
Bones contain protein and calcium and they are strong. But if you knock them sharply or fall awkwardly they can break. Many old people have osteoporosis and their bones break easily. You shouldn't move someone who might have a broken bone as it could damage it more, especially the spine.

↓ A*

Examiner: A clear answer using lots of technical terms properly. Each part of the question has been answered correctly. The answer is easy to follow and the spelling, punctuation, and grammar are good.

Exam-style questions

1 Here is a list of body organs:

muscle   heart   kidney   skin   bone

A01 **a** Which organ contains hair follicles?

A01 **b** Which organ receives blood from the coronary artery?

A01 **c** Which two organs may be joined together by a tendon?

A01 **d** Which organs carry out excretion?

2 Some women find it difficult to become pregnant. They may use IVF, which involves taking eggs from the woman just before ovulation.

A01 **a** What does IVF stand for?

A01 **b** Which organ in a woman's body makes eggs?

A01 **c** What normally happens in a woman's body at ovulation?

A01 **d** During IVF, what is added to an egg to fertilise it?

3 **a** Which gas is excreted by the lungs?

A01

A02 **b** What happens in the lungs to cause an asthma attack?

4 In 1901, Dr Landsteiner discovered blood groups. He said there were three. Scientists now know that there are four blood groups.

A01 **a** Name the four blood groups.

A02 **b** Which group is safe for anyone?

A01 **c** Some people have an inherited condition that means their blood does not clot properly. What is the name of this condition?

A01 **d** Explain the difference between a single and a double circulatory system.

5 The graph gives information on organ transplants between 1996 and 2005.

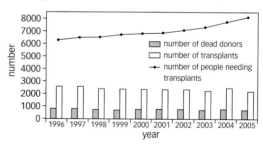

A02 **a** Describe the general trends over the 9 year period shown by the graph.

A02 **b** Describe the 'opt out' organ donation system that could be adopted to boost organ donation.

6 Bypass surgery can treat blocked coronary arteries using a piece of vein from a patient's arm or leg. Explain

A02 why blocked coronary arteries have to be bypassed.

Extended Writing

7 Describe how single-celled organisms,

A01 earthworms, frogs, fish, and humans carry out gaseous exchange.

8 Describe the causes and symptoms of

A01 lung cancer and asthma. How can an asthma attack be treated?

9 Explain how the small intestine is well

A01 adapted for digestion of food and absorption of the products of digestion.

A01   Recall the science

A02   Apply your knowledge

A03   Evaluate and analyse the evidence

B6

Beyond the microscope

Why study this module?

As soon as biologists had the microscope as a basic laboratory instrument, a whole new micro world opened up for us. As microscopes have become more precise, so has our understanding of a new branch of biology called microbiology. From this has come the new scientific field of biotechnology.

In this module you will study the amazing world of the microbe, learning about types of microbes – bacteria, viruses, and fungi. These microbes can be either harmful or helpful. Harmful microbes can cause disease. You will study the ways in which diseases can be transmitted, and their treatment with antibiotics. At the same time, you will consider the early work of some of the pioneers of microbiology. The uses of microbes in the brewing and biofuel industries will also be explored.

You will enter the microscopic worlds of soil and water, investigating some of the impacts of human pollution. Some of the more high-tech uses of microbes will be reviewed, such as the use of microbial enzymes in the food, medical, and detergent industries. Finally, you will examine the use of bacteria or their enzymes in the biotechnological techniques of genetic engineering and DNA fingerprinting.

You should remember

1 The structure of microbial cells.

2 Harmful and helpful microbes.

3 How enzymes work.

4 The action of genes.

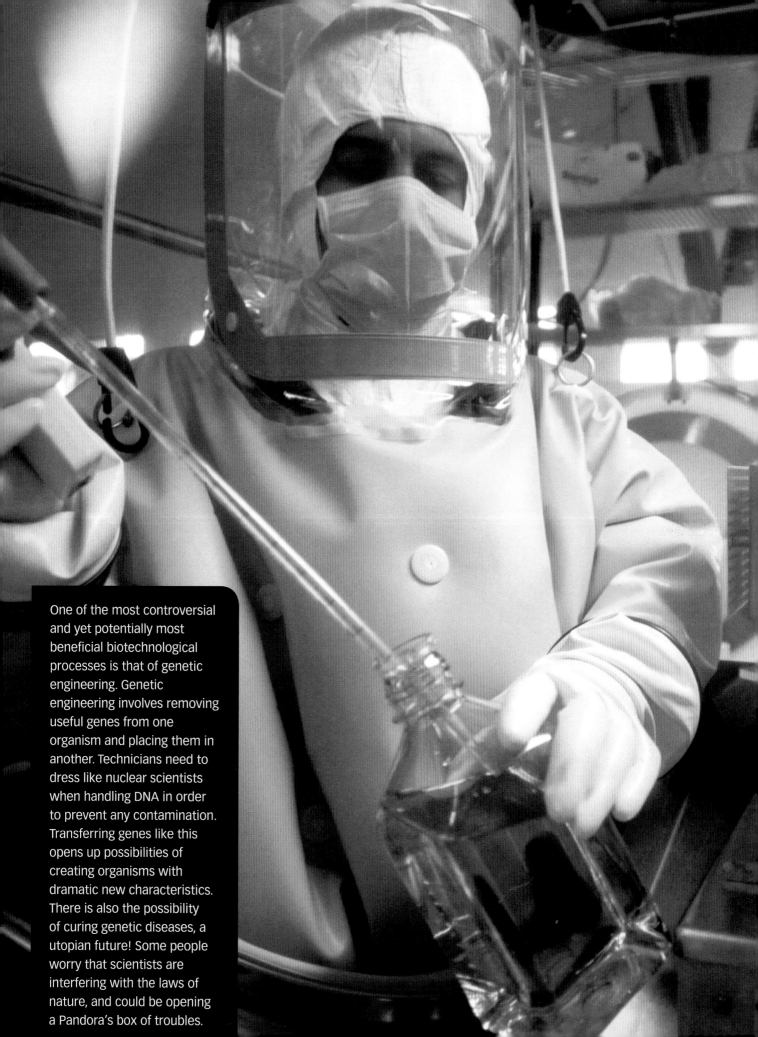

One of the most controversial and yet potentially most beneficial biotechnological processes is that of genetic engineering. Genetic engineering involves removing useful genes from one organism and placing them in another. Technicians need to dress like nuclear scientists when handling DNA in order to prevent any contamination. Transferring genes like this opens up possibilities of creating organisms with dramatic new characteristics. There is also the possibility of curing genetic diseases, a utopian future! Some people worry that scientists are interfering with the laws of nature, and could be opening a Pandora's box of troubles.

Learning objectives

After studying this topic, you should be able to:

✔ know the structure of bacterial cells

✔ recognise that bacteria occupy a wide range of habitats

✔ understand that bacteria reproduce using binary fission

✔ understand the safe handling of bacteria

cocci
(spherical)

▲ Cocci bacteria (*Staphylococcus aureus*) which cause acne spots (× 7000)

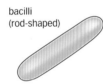
bacilli
(rod-shaped)

▲ Bacillus bacteria (*Escherichia coli*) which can cause food poisoning (× 3000)

vibrio
(curved)

▲ Vibrio bacteria (*Vibrio cholerae*) which cause the disease cholera (× 3500)

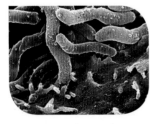

spirilli
(spiral)

▲ Spiral bacteria (*Helicobacter pylori*); which cause stomach ulcers (× 12 000)

Life under the microscope

There is a wide range of organisms that can only be seen using microscopes. Some of these microscopic organisms, or microbes, are helpful to humans; others are harmful. There are three main groups – bacteria, fungi, and viruses.

Bacteria

The microbes that make up the **bacteria** kingdom are single-celled organisms. They are extremely small – a typical bacterium is only a few **microns**, or a few thousandths of a millimetre, long. Bacteria are very important to us. There are many different bacteria, and we can tell them apart and classify them by their shape (see left).

Bacterial cells

At first sight bacterial cells look quite simple. However, they carry out all the functions of other cells.

Bacterial cells can only just be seen using a light microscope. To see fine detail, biologists use high powered microscopes called electron microscopes. These microscopes magnify thousands of times more than a light microscope.

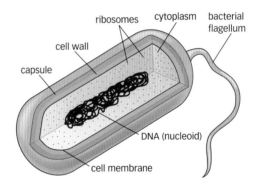

◀ A typical bacterial cell

ribosomes   cytoplasm   bacterial flagellum
cell wall
capsule
DNA (nucleoid)
cell membrane

The electron microscope reveals the following features of bacterial cells:

• cell membrane, controlling the movement of molecules into and out of the cell

• cytoplasm, a jelly-like substance where most of the cell's reactions occur

• cell wall, having the same function as in a plant cell of maintaining the shape of the cell and preventing it from bursting, but made of a different chemical instead of cellulose

• loop of DNA, controlling the cell and its replication. Bacterial cells do not have a nucleus.

- capsule, a slimy protective capsule around the outside of the cell wall in some bacteria. It is this capsule that protects bacteria against antibiotics
- flagellum, whip-like flagella for movement in some bacteria.

Reproduction in bacteria

Bacteria can reproduce very fast. Most of the time they reproduce by dividing into two, and they can do this as quickly as once every two hours or faster. This type of division is called binary fission, an example of asexual reproduction.

An unfortunate consequence of this is that harmful bacteria, such as those that spoil food, can reproduce fast. So food such as milk left out in a warm room will go off very quickly. If you then eat this food, the bacteria may reproduce very quickly in your body. The rate of reproduction may be too fast for your immune system to handle, making you ill.

Bacterial habitats

Bacteria have such a wide range of adaptations that they are found living in almost all environments on Earth. They can live from the depths of the ocean up to the highest mountain peaks. They span cold arctic wastes to volcanic areas and hot springs.

Bacteria survive by obtaining energy from a wide range of sources – some from the Sun by photosynthesis, but others from dead bodies, and from chemical reactions in their cells.

Growing bacteria in the lab

To study bacteria in the lab, they are grown on a jelly called agar in a plate-like dish called a Petri dish. The plates are incubated to keep them warm, and the bacteria grow very fast.

Whenever you grow bacteria you need to take care not to contaminate the plates with other bacteria, and not to allow the bacteria to infect yourself. To do this you need to keep instruments and surfaces free of microbes, or **sterile**. Working in this way is called aseptic or sterile technique.

When bacteria are grown commercially to make a product, they are grown in huge numbers. This is done in a large tank called a fermenter. Again, sterile technique is important.

Key words

bacteria, micron, sterile

A Why do biologists need powerful electron microscopes to study bacterial cells?

B Which part of a bacterium performs the function of a nucleus?

Exam tip

✓ When making a list of the parts of a bacterial cell, focus on the parts that are not found in a plant or animal cell.

Did you know...?

There are bacteria that can survive very high temperatures. They have been found living in environments at temperatures above 80 °C.

Questions

1 Describe the common shapes of bacteria.
2 What is the function of the flagellum?
3 How do bacteria reproduce?
4 Explain why sterile technique is important.

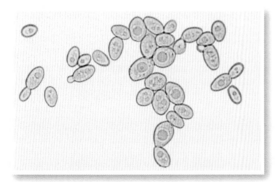

▲ Baker's yeast seen under a powerful light microscope (× 1000)

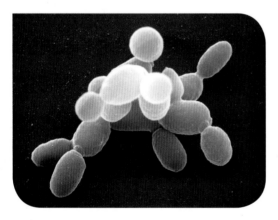

▲ A group of yeast cells all budding from one another

Yeast growth rate

For every 10 °C rise in temperature, the growth rate of yeast doubles. This is only true up to an optimum temperature, above which the yeast's enzymes begin to be damaged.

Fantastic fungi

Fungi are another important kingdom of organisms. They include mushrooms, moulds, and importantly yeasts. Yeasts are commercially useful to us in the making of bread and beers.

Yeasts are single celled, but larger than a bacterial cell. They can be clearly seen under a light microscope, but the internal detail can be seen better using an electron microscope.

The features of a fungal cell

Fungal cells have many parts in common with other cells. The fungal cell has a membrane, cytoplasm, and nucleus which function as they do in plant cells. The cell wall is similar to a plant or bacterial cell. It has the same function, but is made of a different chemical called chitin.

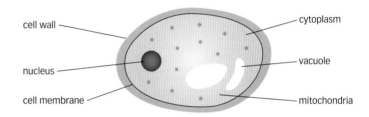

▲ A typical fungal cell

Reproduction in yeast

Yeast cells reproduce mainly asexually by a process called budding. The nucleus divides first, then a bulge forms on the side of the parent cell, which will develop into a new cell. Often the cells remain joined.

Budding can be a fast process. Like binary fission in bacteria, it allows the population of cells to increase rapidly. The optimum growth rate is controlled by

- availability of food
- pH
- temperature
- amount of waste products.

A Name three types of fungi.

B State one way in which a fungal cell and a bacterial cell are: (a) similar (b) different.

C Explain what budding is in yeast.

Viruses

One fascinating group of organisms is the **viruses**. Viruses are much smaller than bacteria or fungi. They are so small they were not seen until the electron microscope was invented. They are about one-hundredth the size of a bacterium.

Viruses do not have a cell structure, and they do not carry out many of the processes of living things. For these reasons, some biologists don't consider viruses to be living things.

The features of a virus

Viruses are made of a protein coat inside which is the genetic material. They cannot live independently – they can only live inside the cells of another organism, called the **host**. Each virus can only invade specific host cells; for example, animal viruses cannot invade plant cells. However, plant, animal, and bacterial cells all have viruses that can invade them. Once inside a host's cell, the virus takes over the cell.

Key words

fungi, viruses, host

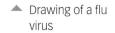
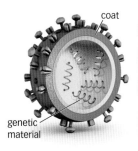
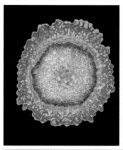

coat

genetic material

▲ Drawing of a flu virus

▲ A flu virus as revealed by an electron microscope (× 300 000)

Reproduction in viruses

The virus takes over the host cell in order to reproduce itself. This reproduction occurs in four main steps.

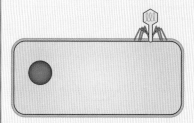
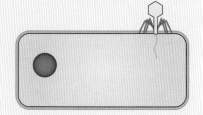

1 The virus attaches to a specific host cell.

2 The genetic material from the virus is injected into the host cell.

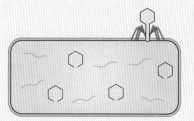

3 The viral genes cause the host cell to make new viruses.

4 The host cell splits open, releasing the new virus.

▲ The four main stages in viral reproduction

Exam tip OCR

✓ If you are asked to compare the reproduction of a fungus and a virus, list the processes and show how they are different.

Questions

1 Arrange the following cells in size order, smallest first: bacteria, virus, yeast. ↓ E

2 Why were viruses not seen until relatively recently?

3 Explain why some biologists think that viruses are not living things. ↓ C

4 Explain why a virus needs a host to reproduce. ↓ A*

Learning objectives

After studying this topic, you should be able to:

- ✔ know about the body's defences against infection
- ✔ know how microbes are transmitted from one person to another
- ✔ understand that major disasters may impact upon the transmission of microbes

Key words

pathogen, contamination, transmission

▲ A cut in the skin forms a potential site for the entry of microbes

Exam tip   OCR

- ✔ Remember that all of the first-line mechanisms are to stop microbes getting into the body. It is the immune system that deals with microbes once inside the body.

Keeping the microbes out

Some microbes, called **pathogens**, cause disease. The body has a number of ways of preventing microbes getting in and causing disease. These features are sometimes referred to as our first line of defence.

Feature	How it prevents entry of microbes	How microbes may overcome the barrier
Skin	Acts as a physical barrier to prevent microbes entering. Washing reduces numbers of microbes on skin. Blood clots at a cut to form a scab and seal the skin.	Cuts in the skin allow microbes in. Insect bites penetrate the skin. Infected needles carry microbes through the skin.
Digestive system (through mouth)	Acid in the stomach kills bacteria.	Eating undercooked food or drinking infected water containing large numbers of microbes.
Respiratory system (through nose)	Cells lining the airways produce a sticky mucus which traps microorganisms. Fine hair-like cilia move the mucus with trapped microbes up to the throat for swallowing.	Some airborne microbes such as cold viruses can get past the cilia. Smoking stops the cilia working.
Reproductive system	Acidic urine kills many microbes.	Some microbes are resistant to acid. Microbes are passed from one person to another by sexual contact.

A List three ways that microbes can enter the body through the skin.

B Describe two ways in which the entry of microbes is prevented in the lungs.

C Why does eating food which is starting to go off make you ill?

Breaking and entering

In order to survive, microbes need to find and enter a host. There are a number of ways that they can get past the host's defences. Here are just a few.

- **Contaminated** food: many bacteria such as *Salmonella* and *E. coli* are common on unwashed vegetables and meat. Food must be washed or cooked correctly to remove or kill bacteria. If not, they will then grow and reproduce. They will then be ingested along with the food. High levels of the bacteria will lead to food poisoning. These bacteria can even spread from uncooked food to cooked food, if people are not careful about food hygiene.
- Contaminated water: cholera is a disease caused by drinking water contaminated with sewage. This water contains the bacterium *Vibrio cholerae*, which causes the disease. Boiling the water kills the bacteria.
- Contact: many microbes are spread or **transmitted** by direct contact with an infected person. The microbe can also be transferred by touching a surface that an infected person has touched. Good hygiene and washing hands after contact will remove bacteria.
- Airborne transmission: viruses like the influenza virus are spread in small water droplets in the air. When someone sneezes, and doesn't cover their mouth, the droplets are fired out into the air for someone else to breathe in.

Natural disasters can spread disease

Natural disasters such as volcanic eruptions and earthquakes can disrupt the systems that prevent the spread of diseases. Cholera and food poisoning may spread easily because

- sewage pipes may be broken, causing sewage to leak out
- water supply systems may be damaged, cutting off the supply of clean fresh water
- electricity may be cut off, so food cannot be refrigerated
- many people may lose their homes and live crowded together in camps, where disease can spread
- the health service may become over-stretched, and lacking in supplies.

> **D** Explain why it is important to wash your hands after sneezing.

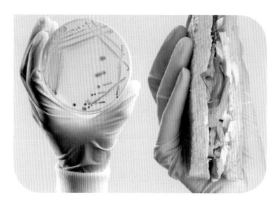

▲ Samples from food can be cultured on an agar plate to test for *E. coli* in the food

◀ Athlete's foot is a common fungal disease spread by direct contact

▲ An open well can easily become polluted, leading to cholera

▲ Sneezing fills the air with cold or flu viruses

Questions

1 Explain how the boy in the picture on the previous page can prevent his knee from becoming infected. **E**

2 Explain why colds and flu may easily be transmitted on a crowded bus.

3 What do you think would be the main priorities to prevent disease after a major disaster? **C**

Learning objectives

After studying this topic, you should be able to:

✔ know that microbes cause many diseases

✔ know the stages of an infectious disease

Key words

disease, **symptom**, **toxin**

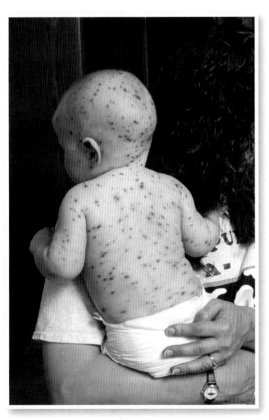

▲ This baby is suffering from chickenpox, a viral disease

Diseases and their microbes

Disease is a state in which the body is not healthy. There are many different diseases, with different causes and different symptoms. Some microbes can cause disease, and these disease-causing microbes are called a pathogen. If the pathogen spreads rapidly from one person to another, the disease is described as infectious. There are pathogens in all the groups of microbes – bacteria, viruses, and fungi.

Bacterial diseases

The table shows some examples of diseases caused by bacterial pathogens. The **symptoms** of a disease are the effects that a patient feels.

Disease	Pathogen and means of transmission
Cholera	Caused by a *Vibrio* bacterium transmitted in contaminated drinking water. The symptoms of the disease are severe diarrhoea and vomiting which lead to dehydration. Cholera is often fatal and develops particularly quickly in children.
Food poisoning	Bacteria are ingested in contaminated food. The symptoms include stomach pains, diarrhoea, and vomiting. It can be so severe that it can be fatal in children and elderly people.

Viral diseases

Here are some examples of viral diseases.

Disease	Pathogen and means of transmission
Influenza	The influenza virus is usually breathed in. Flu is one of the most common diseases. The symptoms include headaches, a running nose, coughs, and sneezes. The disease can occasionally be fatal in weak and elderly people.
Chickenpox	This disease spreads by direct contact, or by breathing in viruses transmitted by an infected person coughing. Symptoms include a rash on the skin that becomes very itchy. Headaches and fevers are also common. Chickenpox is not usually fatal.

Fungal diseases

Fungal diseases include:

Disease	Pathogen and means of transmission
Athlete's foot	The fungus infects the skin between the toes because it is often moist here. The symptoms include cracked flaking skin. It can be painful and in severe cases may bleed.

How an infectious disease develops

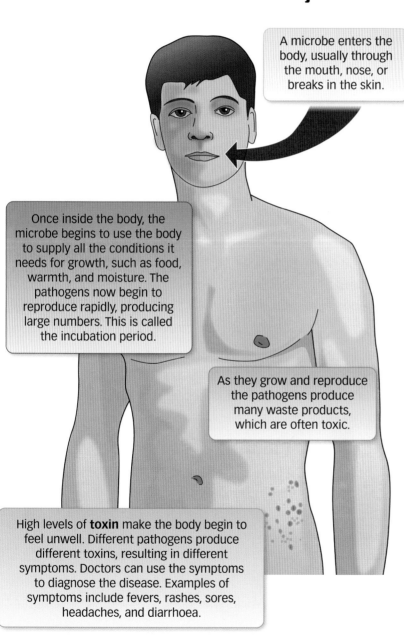

A microbe enters the body, usually through the mouth, nose, or breaks in the skin.

Once inside the body, the microbe begins to use the body to supply all the conditions it needs for growth, such as food, warmth, and moisture. The pathogens now begin to reproduce rapidly, producing large numbers. This is called the incubation period.

As they grow and reproduce the pathogens produce many waste products, which are often toxic.

High levels of **toxin** make the body begin to feel unwell. Different pathogens produce different toxins, resulting in different symptoms. Doctors can use the symptoms to diagnose the disease. Examples of symptoms include fevers, rashes, sores, headaches, and diarrhoea.

▲ Once inside your body, a pathogen reproduces and causes the symptoms of disease

A What is a pathogen?

B What are the symptoms of cholera?

C What makes cholera a particularly serious disease?

Did you know...?

Epidemiology is the study of the numbers and spread of diseases like flu. The data provides information about how the disease is spreading and where it is likely to occur. Governments can plan for outbreaks.

Questions

1 How do we get food poisoning?

2 Why is heart disease not termed an infectious disease?

3 Describe how a pathogen produces its symptoms.

4 Explain why you do not feel ill as soon as a microbe enters your body.

5 Explain how a doctor might use a person's symptoms to diagnose a disease.

Learning objectives

After studying this topic, you should be able to:

✔ know about the work of Pasteur, Lister, and Fleming in treating infectious diseases

✔ understand the development of antibiotic resistance

▲ Louis Pasteur

▲ Joseph Lister

Did you know...?

One of the world's most famous microbiology institutes is named after Louis Pasteur. His body is buried underneath the Pasteur Institute in Paris.

◀ Sir Alexander Fleming

It's a mystery

Until the 1800s, people did not know about microbes. They thought that food decayed because moulds would spontaneously generate on food. Diseases were explained as the effects of evil spirits, or caused by bad smells. There are three great microbiologists who revolutionised our understanding of infectious diseases.

Louis Pasteur

A French scientist called Louis Pasteur (1822–95) proved that decay was caused by microorganisms in the air. He went on to explain that microbes entering the body would cause disease. He proposed the idea that if we could stop microbes entering the body, we could prevent illness. These ideas are known as the germ theory.

Joseph Lister

Armed with the knowledge that microbes entering the body caused illness, Joseph Lister (1827–1912) developed the idea of **antiseptics**. These are solutions that kill microbes. Lister was a surgeon, and he sprayed his instruments with a solution of carbolic acid. This killed microbes on the instruments, which greatly reduced the number of postoperative infections.

Today, many types of antiseptic are used. They are much safer than the acids used by Lister. We use them to kill bacteria on all types of surfaces and on our skin. This contains the spread of microbes, and greatly reduces the number of infections.

Alexander Fleming

More recently, Alexander Fleming (1881–1955) worked in St Mary's Hospital in London. He discovered that a mould called penicillin produced a chemical that would kill bacteria. The fungus grew on one of his agar plates of bacteria. It caused an area where the bacteria could not grow, as they were killed by the penicillium.

During the Second World War scientists were able to make sufficient penicillin to give to a patient, and the patient recovered from the illness. Penicillin was the drug that killed bacteria – the first **antibiotic**.

A Who proposed that microbes caused disease?

B Explain how Lister made surgical procedures safer.

C Explain why the discovery of penicillin was so important.

Key words

antiseptic, antibiotic, resistance

Resistance to antibiotics

Antibiotics were regarded as wonder drugs. However, they cannot cure all infectious diseases:

- Antibiotics do not kill viruses. This is because viruses do not feed, and do not have a cell structure to damage. These are the two main ways that antibiotics work on bacteria.
- Some bacteria can develop **resistance** to an antibiotic, and the drug no longer works on them.

How bacteria develop resistance

The development of antibiotic resistance by bacteria is one of the best examples of evolution by natural selection.

- A mutation occurs in some bacteria, which gives them resistance to the antibiotic.
- Treatment by the antibiotic kills the bacteria in the population that do not have this mutation, so are not resistant.
- The bacteria with the resistance survive.
- The surviving bacteria reproduce, passing the resistance gene on.
- Eventually, the whole population becomes resistant.

Modern doctors are very aware of the problems of antibiotic resistance. They have changed their use of antibiotics over the last 20 years. There are two main practices that they now follow:

1. Doctors only prescribe antibiotics when really necessary. They do not use them for viral conditions or minor illnesses. This reduces the chance of antibiotic-resistant bacteria becoming the most common strain.
2. Patients are encouraged to complete the course of any antibiotics that they are given. This way all the microbes should be killed before resistance can fully develop.

Exam tip OCR

✔ You need to remember the names of Pasteur, Lister, and Fleming, what they did, and how it improved our understanding of microbes and diseases.

Questions

1 Before Pasteur, what did people think caused infectious diseases?

2 Explain why it was important that Lister was aware of the work of other scientists like Pasteur, before he made his discoveries.

3 Why do you think the original penicillin discovered by Fleming is virtually useless today?

4 Explain why modern biologists need to carry out large numbers of clinical trials on any new antibiotic, before it can be used by doctors.

Key words

fermentation

▲ These dairy products are made using processes that depend on bacteria

> **A** Name three useful products made by bacteria.
>
> **B** Explain why cheese and yoghurt are important in the diet.

Microbes in industry

Microbes are not all bad. Both bacteria and fungi can be used by humans to carry out useful tasks, such as to make a useful product on both a domestic and a commercial scale.

Products from bacteria

Bacteria have been used in a variety of ways for centuries. However, with our greater understanding of bacteria modern biologists have been able to make more efficient use of them. Today bacteria are used in the manufacture of:

• Yoghurt: this popular dairy product made for over 5000 years. It is made by adding the bacterium *Lactobacillus* to milk. Yoghurt is a very nutritious food which is rich in protein, calcium, and vitamins.

• Cheese: another common dairy product made for about 5000–8000 years. Cheese is made by causing milk to curdle, or separate into a solid curd and liquid whey. Curdling can be achieved using a mix of enzymes and bacteria such as *Lactobacillus*. The solid part of the milk is then turned into cheese. Like yoghurt, cheese is rich in protein, calcium, and vitamins.

• Vinegar: used for at least 5000 years, it was even recorded in Egyptian times. It is produced by the acidifying of wine, cider, or beer to produce wine, cider, and malt vinegars. The production of vinegar uses bacteria such as *Acetobacter*.

▲ Vinegars are made by the action of *Acetobacter* on wine, cider, or beer

- Silage: a common winter fodder for cattle, which has been in use since the 1880s. Green cut vegetation is piled up in a large heap and covered in plastic, or placed in a silo. The vegetation is broken down by **fermentation**, or anaerobic respiration. The process will occur naturally, but it is speeded up by adding another *Lactobacillus* species.
- Composting: a natural process used for at least 2000 years. Since the 1920s it has become important in organic farming. The dead remains of plants and animals are digested by bacteria and fungi to form nutrient-rich soils. Often the remains are simply piled up at the end of the harvest and allowed to rot down until the next planting season.

▲ Silage bales in Wales

> C How is silage important to farmers?

A closer look at yoghurt-making

The yoghurt industry is worth millions of pounds a year. Making yoghurt involves adding bacteria to milk. Like all processes involving microbes, it is important that all equipment is sterile – clean and free from microbes – throughout the process.

▲ Quality control sampling in a yoghurt-making factory

▲ Colours and flavours are added to make the final product

Exam tip | OCR

✔ Remember the process of yoghurt-making as a sequence of steps.

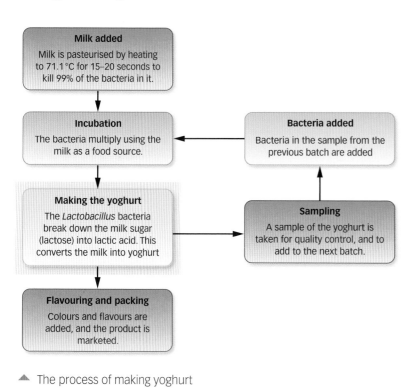

Milk added
Milk is pasteurised by heating to 71.1°C for 15–20 seconds to kill 99% of the bacteria in it.

Incubation
The bacteria multiply using the milk as a food source.

Bacteria added
Bacteria in the sample from the previous batch are added

Making the yoghurt
The *Lactobacillus* bacteria break down the milk sugar (lactose) into lactic acid. This converts the milk into yoghurt

Sampling
A sample of the yoghurt is taken for quality control, and to add to the next batch.

Flavouring and packing
Colours and flavours are added, and the product is marketed.

▲ The process of making yoghurt

Questions

1 Why are fruit and sugar added to many yoghurts?

2 Explain why sterile technique is important when producing yoghurt.

3 Explain why a sample from a previous batch is added at the start of the yoghurt-making process.

E

C

Useful fungi

As well as bacteria, fungi are also used to make useful products commercially. Yeast is probably the most widely used fungus. Among other products, fungi are used to make alcoholic drinks and bread.

Brewing

Brewing is the production of alcoholic drinks by the process of fermentation. Yeast makes the alcohol by fermenting sugars found in plants, usually in the fruit or grain (seed).

- Wine is made from grapes.
- Beer and lager are made from malted barley.
- Cider is made from apples.

Step	Process	Explanation
1.	Malting	This happens in beer- and lager-making. The barley seeds start to germinate (grow), which converts the starch stored in the seeds into sugar.
2.	Extracting the sugar	The plant material is either crushed or soaked in water to get the sugar out.
3.	Flavouring	Wines get their flavour from the fruit juice used, while beers are flavoured with hops, the female flowers of the hop plant, *Humulus lupulus*.
4.	Adding yeast	The container is sealed to prevent air and other microbes entering. The yeast can respire aerobically for a very short time, until any oxygen in the container is used up. This allows the yeast to reproduce. The liquid is kept warm.
5.	Fermentation	The culture quickly becomes anaerobic and aerobic respiration does not happen; the yeast starts to produce alcohol. If there is too much oxygen in the container, anaerobic respiration does not occur and vinegar is produced. This process happens in large stainless steel vats called fermenters. The mixture is kept at a constant temperature of 25–30 °C, which gives the best rate of respiration to produce alcohol. This temperature also gives a better flavour.
6.	Extracting the wine and beer	After fermentation is over, the liquid is separated from the yeast cells. Usually the yeast is allowed to sink to the bottom to separate it out of the liquid. The liquid may need clarifying by a filtration process.
7.	Pasteurising and packaging/bottling	The product is heated and quickly cooled to kill any remaining microbes. This gives the product a longer shelf life when stored in bottles.

The fermentation reaction

This is the anaerobic respiration reaction (without oxygen) that occurs in the yeast cells:

glucose (sugar) → ethanol (alcohol) + carbon dioxide

Symbol equation for the fermentation reaction

Here is the symbol equation for anaerobic respiration in yeast: $C_6H_{12}O_6 \rightarrow 2C_2H_5OH + 2CO_2$

Making it stronger

There is a limit to the concentration of alcohol that can be produced by the brewing process. Alcohol is toxic and eventually kills the yeast in the fermenter when it reaches a certain concentration. Some yeasts can tolerate more alcohol than others. This results in drinks with different alcohol contents:

- Beers and lagers usually contain 3–5% alcohol.
- Wines usually contain 11–12% alcohol.

Spirits such as vodka and whiskey contain high levels of alcohol. To make spirits, the first step is fermenting plant material, similar to the production of beers and wines.

- Rum is made from sugar cane.
- Whiskey is made from malted barley.
- Vodka is made from potatoes.

Spirits typically contain 40% alcohol. To increase the concentration of alcohol from the fermentation process, the liquid is **distilled**. To do this the liquid is placed in a large container, or still, and heated. The alcohol boils at a temperature of about 80°C, which is lower than the boiling point of water. The alcohol rises up the column as a vapour, leaving the water behind. The vaporised alcohol passes along a collecting arm, and cools. This product contains a lot more alcohol than the fermentation liquid.

◀ Stills at a whiskey distillery. Alcohol is distilled off the fermentation liquid to produce a spirit with a higher alcohol content. Distillation can only be carried out in premises licensed for the production of alcohol.

Key words

brewing, distillation

Did you know...?

Food processing factories produce waste water containing sugars. Yeast can be used to ferment the sugars and clean the waste water.

Exam tip   OCR

✔ The fermentation process is a sequence of steps. Remember the differences between brewing and making yoghurt.

A Which microbe is involved in the making of alcohol?

B Name the two products of alcoholic fermentation.

C Why are hops added to beer?

Questions

1 Name two alcoholic drinks produced by fermentation. ↓ E

2 Why do spirits keep longer than beers and wines? ↓ C

3 Explain why fermentation alone does not produce spirits.

4 Explain why the temperature must be carefully regulated:
(a) in the brewing process
(b) in the distilling process. ↓ A*

Key words

biofuel

A What is the source of energy for making biofuels?

B Name three types of biofuel.

Balancing the books

To burn fuels while maintaining no overall increase in greenhouse gases is a difficult balancing act. When we burn biofuels we have grown, the carbon dioxide taken in during photosynthesis is then released during the combustion.

However, land is needed to grow these crops. In some areas forests are cleared for the cash crop and this leads to a loss of plants to absorb carbon dioxide, and an increase in carbon dioxide released by decaying wood. It also causes a loss of habitat and extinction of species.

Greener fuels

The burning of fossil fuels harms the environment as it produces waste gases including carbon dioxide, which leads to global warming. A variety of fuels from biological materials can be used as an alternative. These are called **biofuels**. They are better for the environment because the carbon dioxide produced when they burn is balanced by the carbon dioxide they use in photosynthesis while they are growing.

What are biofuels?

As in fossil fuels, the energy in biofuels originates from sunlight used in photosynthesis. Photosynthesis produces the biomass in plants, and this biomass can be used directly or indirectly as biofuel. Wood can be burnt directly to release energy. Fast-growing trees can be used to fire power stations.

Common biofuels include wood, biogas, and alcohol.

Advantages of using biofuels	Disadvantages of using biofuels
Reduce fossil fuel consumption by providing an alternative fuel.	Cause habitat loss because large areas of land are needed to grow the plants.
No overall increase in levels of greenhouse gases, as the plants take in carbon dioxide to grow, and release it when burnt.	Habitat loss can lead to extinction of species.
Burning biogas and alcohol produces no particulates (smoke).	Data shows that some biofuels transfer less energy than other fuel types.

Biogas

Biogas is made by the fermentation of carbohydrates in plant material and sewage by bacteria. This fermentation occurs naturally, for example in marshes, septic tanks, and even inside animals' guts. Biogas is also produced at some landfill sites, where the gas can be burnt. Sometimes the biogas can explode, making the landfill site unusable for many years.

Biogas is a mixture of gases that will burn in oxygen, forming a useful fuel:

- methane (50–75%)
- carbon dioxide (25–50%)
- hydrogen, nitrogen, and hydrogen sulfide (less than 10%).

Small-scale biogas production

In remote areas of Nepal and India, lacking mains electricity and mains sewage systems, biogas is made for families and used for cooking. The fermentation happens in a large tank sunk into the ground, which keeps the temperature constant. The family place organic material like dead plants and animal waste in the tank. The bacteria digest this waste, releasing the gas.

Biogas production on a larger scale

The gas is generated commercially in large anaerobic tanks. Wet plant waste or animal manure continuously flows, and the gas produced is removed. The remaining solids need to be removed from the tanks and can be used as a fertiliser in some cases. Gas production is fastest at a temperature of 32–35 °C, because the fermenting bacteria grow best at this temperature. At these temperatures the bacterial enzymes work best. Any higher would denature the enzymes.

Biogas has a number of uses:
- as vehicle fuel
- to generate electricity
- for heating systems.

Bioethanol

Alcohol is produced from plant material by yeasts in brewing. On a larger scale this alcohol can be used as a fuel. Mixed with petrol it produces gasohol, which is a common fuel for cars. This is a particularly economic fuel in countries that produce large amounts of plant waste, such as Brazil. Brazil has no oil reserves and plenty of sugar cane waste to make the alcohol.

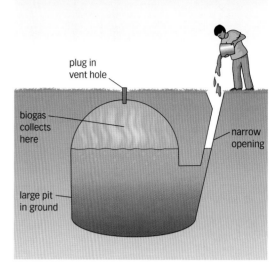

▲ Section through a biogas digester

Problems with biogas

There are a few technical issues with the production of biogas. First, since many different waste materials are used, a large range of bacteria are needed to digest the waste.

Biogas is a cleaner fuel than petrol or diesel, as fewer particulates are released. However, burning biogas releases 4.5–8.5 kWh/m³ of energy compared with natural gas, which releases 9.8 kWh/m³. This is because biogas contains less methane than natural gas.

A final difficulty is that if the biogas becomes mixed with air, so that there is more oxygen and the methane content drops to 5–20%, the mixture becomes explosive. This is not a problem when the gas is contained and not allowed to mix with the air. Concentrations above 50% can be burned in a controlled way.

Questions

1 Explain why biogas from landfill sites is particularly dangerous.

2 Give two reasons why gasohol is used in Brazil.

3 Why must a biogas digester be kept airtight?

4 Explain why using biofuels should not contribute to any net increase in greenhouse gases, in contrast to using fossil fuels.

Learning objectives

After studying this topic, you should be able to:

✔ know that the land surface is covered in rock or soil

✔ know the composition of soil

✔ describe organisms that live in the soil

✔ understand the importance of the earthworm

Key words

humus

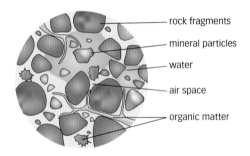

- rock fragments
- mineral particles
- water
- air space
- organic matter

▲ The constituents of soil

Soil constituent	How to test a soil sample for it
moisture	Weigh, bake, and reweigh sample
humus	Weigh, burn, and reweigh sample
air	Dry, weigh the sample and measure its volume, calculate density

A Name three types of soil.

B Which type of soil retains water best?

Solid ground

The land surface is covered in either bare rock or soil. Rock is weathered to form the soil.

What is soil?

Soil contains a number of different components:

- Fragments of rock (minerals) – these are produced when rock is weathered. The size varies, and this determines the type of soil.
- Air spaces – gaps between the particles.
- Water – this fills some of the spaces between particles.
- Dead material – fragments of dead plants, animals, or organic waste.
- Living organisms – there is a huge variety of life in and on the soil. Plants rely on the soil for minerals, water, and to anchor them.

Soils have different structures depending on the size of their particles.

Soil type	Particle size	Air spaces	Permeability to water
clay — small clay particles, tiny air spaces	small (less than 0.002 mm)	few and small	low – water is retained in soil; soil can flood
loam — clay and sand particles, many air spaces of different sizes	mixture of small and large	many and variable in size	medium – water retention is good
sand — large sand particles, many large air spaces	large (0.05–2 mm)	many and large	poor – little water retained in soil

Soil as a habitat

Most organisms in the soil need water and also oxygen for respiration. The amounts of water and oxygen in a soil depend on the soil particle size.

If the particles are small (clay soil) there will be few air spaces, and they will often be full of water, reducing oxygen levels.

- If the particles are big (sandy soil) there will be plenty of oxygen in the air spaces, but the water will drain away.

An ideal soil (loam) has a mixture of particle sizes, providing both air spaces and water retention. Gardeners improve their soil by digging to mix the layers, allow in air and to increase drainage.

Dead material decomposes in the soil to produce **humus**. This releases minerals into the soil which are needed by plants for growth. Humus adds a fibrous quality to soil – it tends to hold soil particles apart, improving aeration. It also helps retain water in the soil.

Soil pH also affects what can live there. Some plants such as heathers grow well in acidic soil. Many plants prefer relatively neutral soils. Alkaline soils are rare in the UK. Many farmers add lime to neutralise acidic soils so that they can grow more crops.

What lives in soil?

There is a whole community of organisms living in soil, linked together to form a food web.

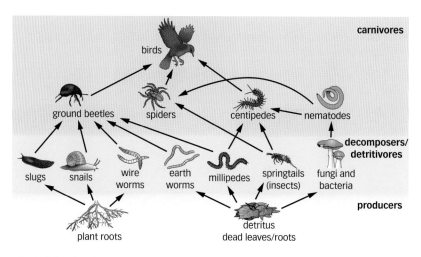

▲ Soil food web

Did you know...?

The famous biologist Charles Darwin was one of the first to study the biology of the earthworm. He recognised how important earthworms are in improving the fertility and structure of the soil.

Worms pull dead leaves down into the soil, burying them. This organic material is then slowly decayed by bacteria and fungi, improving the nutrient content of the soil.

Earthworm burrows create gaps which aerate the soil. As the worms move through the burrows they push air through them. The cavities also allow water to drain more freely, reducing the chance of flooding.

The earthworm mixes the soil by eating soil and passing out waste elsewhere in the soil.

Earthworms release calcium carbonate into their gut to help the digestion of leaves. This then passes out in their waste, and has the added bonus of helping to neutralise acidic soils.

▲ Earthworms as soil improvers

Questions

1. List the main components of soil.

2. Using the food web, explain why soils with little detritus contain few organisms.

3. Detritivores eat detritus. Name three detritivores in the soil food web.

3. Detritivores eat detritus. Name three detritivores in the soil food web.

4. If a gardener used an insecticide that killed ground beetles, what might happen to the earthworm population?

E

C

Learning objectives

After studying this topic, you should be able to:

- ✔ list some advantages and disadvantages of living in water
- ✔ know that phytoplankton are the producers in ocean food webs
- ✔ explain seasonal fluctuations in plankton numbers

Key words

plankton

Water regulation

In fresh water, too much water can enter the body by osmosis. This is not a problem for plants, as cell walls stop cells expanding and prevent excess water getting in. In animals, excess water must be removed from the body. Freshwater fish urinate frequently, removing the excess. Microscopic organisms such as amoebae have a cell structure called the contractile vacuole, into which the excess water goes. The vacuole then moves to the cell surface, fuses with it, and releases the water.

Sea water is salty, and this affects osmosis. Many invertebrates have bodies at the same salt concentration as the sea and so have no problem, but some larger fish do not. They actively get rid of the salt in the water they drink.

Living in water

There is a huge diversity of life in water, including microorganisms.

The wonders of water

There are several advantages of living in water:

- Buoyancy – water is more dense than air, so it gives more support to the organisms that live in it. The largest animal, the blue whale, can measure up to 30 m in length and weigh up to 170 tonnes. These animals could not support their weight on land.
- Removal of waste – animal waste is washed away and does not build up, as it is greatly diluted in the water and broken down.
- Steady temperature – surface waters vary in temperature, but not as much as the air does. Water requires a lot of energy to heat it up. The waters at the poles are at about 0 °C, while temperatures near the Equator can be up to 30 °C. Deeper water has a very stable temperature, about 4 °C. Aquatic organisms do not have to cope with extremes or rapid changes in temperature.
- Ready water supply – living in water means that there is no risk of dehydration.

The woes of water

There are also disadvantages of aquatic habitats:

- Movement – water is more dense than air, making it harder to move through, so aquatic animals use more energy.
- Water balance – water is everywhere; the problem is balancing the amount of water in the body.

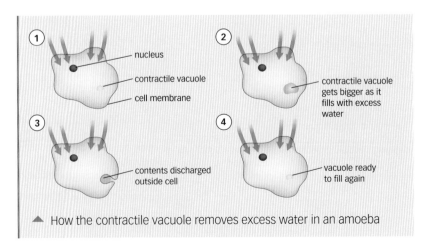

▲ How the contractile vacuole removes excess water in an amoeba

Aquatic food webs

There are microscopic organisms living in water called **plankton**. There are two types:

- Phytoplankton are photosynthetic microorganisms – they are producers.
- Zooplankton are animal-like microorganisms – they are consumers.

Plankton float in the open waters, moving in currents. The numbers of the phytoplankton vary during the year. Three factors control their numbers:

Factor	Seasonal effect	Effect of depth
Light	As day length increases, the numbers of phytoplankton increase.	Light only penetrates surface waters, so phytoplankton are limited to the surface waters.
Temperature	As surface water temperatures rise in spring, the numbers of phytoplankton increase.	At depth the temperature is a constant 4 °C, too cold for phytoplankton to grow.
Minerals	Minerals rise to the surface during the winter, so there is a ready supply for phytoplankton in the spring.	Mineral concentrations increase at depth, but phytoplankton are limited there by light and temperature.

Seasonal changes in phytoplankton numbers will influence the numbers of zooplankton.

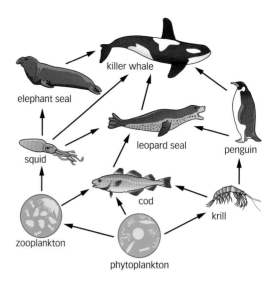

▲ A marine food web

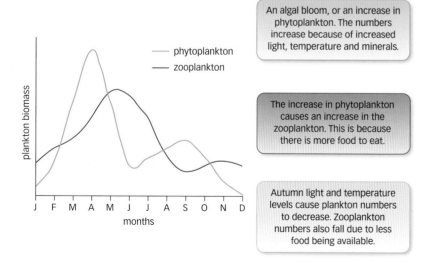

An algal bloom, or an increase in phytoplankton. The numbers increase because of increased light, temperature and minerals.

The increase in phytoplankton causes an increase in the zooplankton. This is because there is more food to eat.

Autumn light and temperature levels cause plankton numbers to decrease. Zooplankton numbers also fall due to less food being available.

▲ Seasonal cycles in plankton populations in the North Atlantic

Aquatic grazing

Grazing food webs, based on photosynthetic producers, are common in the surface layers of the oceans.

In deep water food webs do not start with phytoplankton as there is no light. Food chains at these depths rely on dead food falling from above, called marine snow. These are detrital food chains.

Other food chains start with bacteria – producers that get their energy from chemical reactions in a process called chemosynthesis.

Questions

1. Explain why phytoplankton do not grow in the deep oceans.

2. Why is the phytoplankton population at its peak in April?

3. Explain why the phytoplankton population is low in June, when growing conditions are good.

Learning objectives

After studying this topic, you should be able to:

- ✔ know that organisms are affected by pollution
- ✔ understand the steps in the process of eutrophication
- ✔ appreciate that pollutants can build up in some species

Key words

eutrophication, indicator species

▲ Scientists cleaning a pelican caught in the 2010 oil spill in the Gulf of Mexico. It was estimated that hundreds of thousands of gallons were spilling into the water each day.

What a waste!

Unfortunately, humans generate a lot of waste. Many of these wastes pollute water, affecting the number and type of organisms (including microscopic ones) that live there.

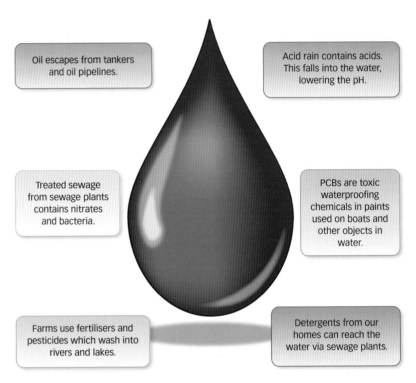

Oil escapes from tankers and oil pipelines.

Acid rain contains acids. This falls into the water, lowering the pH.

Treated sewage from sewage plants contains nitrates and bacteria.

PCBs are toxic waterproofing chemicals in paints used on boats and other objects in water.

Farms use fertilisers and pesticides which wash into rivers and lakes.

Detergents from our homes can reach the water via sewage plants.

▲ Sources of water pollution

A Name two pollutants of water.

B Explain how fertilisers sprayed on crops get into water.

C In 2010 there was a rupture in an oil pipeline off the coast of New Orleans, which released vast quantities of oil into the sea. What is the effect of oil on wildlife?

Eutrophication: a case study

One major pollutant of water is nitrates. Nitrates enter the water in untreated sewage or directly from fertilisers. When this happens the process of **eutrophication** occurs.

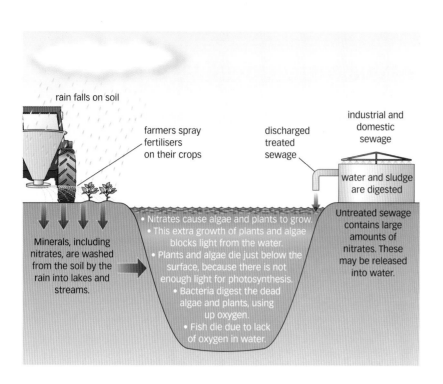

rain falls on soil

farmers spray fertilisers on their crops

discharged treated sewage

industrial and domestic sewage

water and sludge are digested

Minerals, including nitrates, are washed from the soil by the rain into lakes and streams.

• Nitrates cause algae and plants to grow.
• This extra growth of plants and algae blocks light from the water.
• Plants and algae die just below the surface, because there is not enough light for photosynthesis.
• Bacteria digest the dead algae and plants, using up oxygen.
• Fish die due to lack of oxygen in water.

Untreated sewage contains large amounts of nitrates. These may be released into water.

▲ River pollution causing severe algal bloom

Indicators of pollution

Biologists can get an idea of the level of water pollution by looking for the presence of **indicator species**. For example:

- pH changes in water can be indicated by a reduced number of amphibians in polluted streams and bogs.
- Reduced oxygen levels are indicated by rat-tailed maggots.

Poisoned whales

Certain chemicals do not break down in the environment quickly. Examples include commonly used industrial chemicals called PCBs, and the pesticide DDT. You may remember that chemicals like these can build up in food chains. They will eventually reach toxic levels in the top carnivores.

Both of these chemicals can build up in the bodies of whales over many years. The result is that some whales, such as killer whales, are among the most contaminated animals on Earth. PCBs are known to suppress the immune systems of these animals. This may contribute to the decrease in their numbers. Before the 1960s, whaling was common practice in many communities. Humans ate the whale meat. The whale meat was so contaminated that the levels of PCBs had a harmful effect on the people who ate it.

Questions

1 What is the effect of pollution on the numbers of aquatic organisms?

2 Draw a flow diagram to describe eutrophication.

3 Explain how indicator species might suggest reduced oxygen levels in water.

4 Environmentally friendly detergents do not contain as many nitrates and phosphates as mainstream detergents. Explain how this will reduce eutrophication.

5 Explain how PCBs build up in whale meat.

E

C

A*

A Give one reason why enzymes are used in a variety of industries.

B Name two industries that use enzymes, and suggest why they are used.

C Why are different enzymes used in different industries?

Making microbes work for us

Enzymes are catalysts, and in biotechnology scientists often use enzymes to speed up chemical reactions. This use of enzymes in industry is called **enzyme technology**. Bacteria are easy to grow in large quantities, so they are used to produce these enzymes on a large scale.

Industrial uses of enzymes

Industry	Enzyme	Use
dairy (eg cheese making)	protease (eg rennet)	Causes solids (curds) to separate from the liquid (whey) in milk.
food processing	proteases	Digest proteins in foods such as soy and citrus products, to remove bitter tastes.
fruit juice	cellulases	Break down cell walls in the fruit, releasing juice.
	amylases	Reduce cloudiness and increase sweetness by breaking down starch.
medical	glucose oxidase	Present in kits used by people with diabetes to test for glucose in urine or blood.
biological washing powder	proteases, lipases, and amylases	Remove organic stains such as food and grass stains from laundry.

Sweet enough?

The food industry makes great use of enzymes. As a nation we have a very sweet tooth. Processed foods and soft drinks contain a lot of sugar to sweeten the product. The sugar extracted from plants like sugar cane is called sucrose. This is not as sweet to the taste as other sugars like fructose unless it is broken down by enzymes.

An enzyme called invertase (sucrase) digests sucrose into glucose and fructose. The food industry now has a much sweeter product. They can use fructose to sweeten food. This way less sugar is needed to produce a sweet product. This is common practice in producing low calorie foods and drinks.

Cleaning power

Biological washing powders contain soap powder, enzymes, and minerals. Why are the enzymes added?

- Enzymes digest stains.
- Enzymes digest fibres in the dirt, releasing bobbles.
- They allow stain removal to occur at a lower temperature, which saves energy and money.

These washing powders contain several different enzymes, each of which does a different job. The enzymes break down the stains into small soluble products that wash off the fabric. Unfortunately, the enzymes in biological washing powders don't work at high temperatures or extreme pH levels.

Disadvantages of biological washing powders

Biological washing powders allow us to use less energy and to machine wash delicate fabrics, because they do not need high temperatures. However, there are a few limitations to using enzymes in this way:

- The enzymes may be destroyed at high temperatures so do not work for a hot wash.

- The enzymes are pH sensitive, so they may not work so well in areas of the country with particularly acidic or alkaline water.
- They may cause allergies.

▲ Slimming products

Amylases digest carbohydrate stains such as starch from foods like pasta and flour.

Lipases digest fats and fatty stains like grease.

Proteases digest protein stains such as blood.

▲ Biological washing powders contain a combination of enzymes

D Suggest why people might want to reduce their sugar intake.

E Explain why less fructose than sucrose is needed to sweeten food.

Questions

1 Why are enzymes used in cheese making?

2 Explain why biological washing powders are better at removing stains than non-biological powders.

3 Why is it not efficient to use biological washing powders in a hot wash?

Key words

immobilised

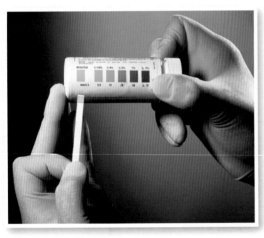

▲ Reagent sticks are used to test for glucose in urine. They contain a range of enzymes.

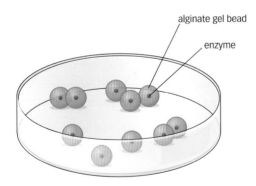

alginate gel bead

enzyme

▲ Alginate beads can be used to immobilise enzymes

Doctor, doctor!

The medical industry makes great use of enzymes. Two examples are:

- Testing for sugar – people with diabetes need to keep a watchful eye on the glucose levels in their urine or blood. The normal blood glucose level is from 4–8 mmol/l. Diabetics often experience higher levels. They monitor their blood glucose level to prevent it rising for too long, as persistent high levels can damage blood vessels in organs like the eyes. If their blood glucose level falls too low, they could become unconscious. Testing could be done using the food test called Benedict's test for glucose. This is not very practical in our modern lives. Now we use reagent sticks to test for glucose. These sticks make use of enzymes.

- Lactose intolerance – some people don't make the enzyme lactase, and so they cannot digest the sugar lactose in their gut. Bacteria in their digestive system ferment it instead, leading to diarrhoea and wind. Enzymes are now used to produce lactose-free foods.

Immobilised enzymes

Enzymes are delicate molecules. They are highly temperature sensitive. This can make them difficult for scientists to use and store. Scientists have developed a method to help make enzymes more stable and easier to use.

The enzymes are **immobilised**, which means they are attached to a more inert substance. There are two common ways of doing this. The enzyme can be added to a fibre mesh, as on reagent sticks. Another method is to produce gel beads containing the enzymes. To do this:

- The enzymes are mixed with a solution of sodium alginate.
- The mixture is dropped into a calcium chloride solution.
- This causes small beads of alginate gel to form.
- Embedded in the gel are the enzymes.

The advantages of the use of immobilised enzymes are:
- The beads are easier to use than free enzymes.
- The beads support the structure of the enzyme, making it less sensitive to temperature and pH.

- The enzyme can be removed from a reaction mixture by filtering out the beads, so the solution is not contaminated with enzymes.
- The beads can be placed in a column, and the reactants poured in at the top. The products can be drawn off at the bottom of the column.

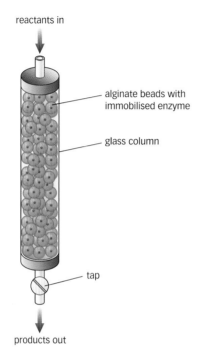

reactants in

alginate beads with immobilised enzyme

glass column

tap

products out

▲ Immobilised enzymes can be used in a column to make a continuous flow of products

Lactose intolerance

Lactose is a sugar found in milk. As mammals, we are all fed on milk as babies. However, a surprisingly large number of people show some signs of lactose intolerance. This means that they are unable to digest the lactose in their diet. If they eat or drink too much lactose, it can make them unwell.

Beads can be made containing immobilised lactase enzymes. These beads are then mixed with high-lactose foods such as milk. The enzyme digests the lactose into glucose and galactose:

$$\text{lactose} \xrightarrow{\text{lactase}} \text{glucose} + \text{galactose}$$

Both glucose and galactose are easily absorbed in the gut.

A State two medical uses of enzymes.

B Explain why reagent sticks are easier to use to test for glucose than the Benedict's test.

C Explain why it is easier to remove immobilised enzymes from a reaction.

Did you know...?

Nearly all cats are lactose intolerant, so the traditional idea of giving cats a saucer of milk is actually bad for them. Special cat milk is now available, which has had the lactose removed using immobilised enzymes.

◀ Normal milk is bad for cats

Questions

1 What are the two ways in which scientists can immobilise enzymes? ↓ E

2 Describe how large sucrose could be broken down using a column of beads.

3 Explain why using immobilised enzymes in bead columns is more economic than using free enzymes. ↓ C

4 Explain why it is important for the alginate beads to be porous. ↓ A*

A Name two products produced by genetic engineering.

Productive bacteria

Biotechnology is a growing industry. It develops ways of using microorganisms for industrial processes.

One such process is called **genetic engineering**. It is possible to introduce a gene from one organism (the donor) into bacteria (the host), so that the bacteria will make useful protein products for us. Two well established examples of products made in this way are

- insulin: used by people with diabetes to control their blood sugar
- human growth hormone: used to treat people with reduced growth.

The process of genetic engineering involves these basic steps:

1. Identifying the gene that codes for the protein to be produced.
2. Removing it from the donor, e.g. a human cell.
3. Introducing it into a host, the bacterium.
4. Growing the bacteria on a large scale to make the product.

Engineering bacteria
Producing transgenic bacteria

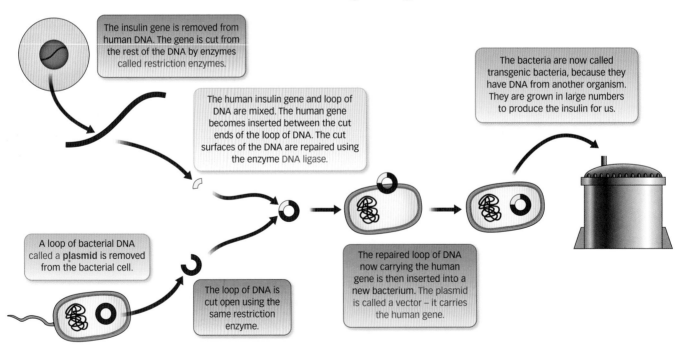

The insulin gene is removed from human DNA. The gene is cut from the rest of the DNA by enzymes called restriction enzymes.

The human insulin gene and loop of DNA are mixed. The human gene becomes inserted between the cut ends of the loop of DNA. The cut surfaces of the DNA are repaired using the enzyme DNA ligase.

The bacteria are now called transgenic bacteria, because they have DNA from another organism. They are grown in large numbers to produce the insulin for us.

A loop of bacterial DNA called a **plasmid** is removed from the bacterial cell.

The loop of DNA is cut open using the same restriction enzyme.

The repaired loop of DNA now carrying the human gene is then inserted into a new bacterium. The plasmid is called a vector – it carries the human gene.

▲ Genetic engineering is used to produce human insulin in large quantities. Before this process was developed people with diabetes injected insulin from cows or pigs.

Cloning the bacteria

Once the plasmid has been taken up by the bacteria, these **transgenic** bacteria need to multiply to produce a large culture of bacteria, all capable of making insulin. The bacteria are placed in a large container called a fermenter. Here ideal conditions can be provided for their rapid growth and reproduction. The bacteria divide asexually, producing genetically identical copies of themselves. This is called cloning. The resulting large culture of bacteria will make large amounts of the insulin protein. This is harvested and packaged.

Checking for the gene

Genetic engineering can be tricky. Biotechnologists can't see the actual genes, and sometimes the bacteria will not take up the plasmid with the gene. These bacteria will not make the protein, so there is no point in culturing them. Biotechnologists test the bacteria to see if the gene is there. This testing uses a process called an assaying technique.

The assaying is usually done by growing the bacteria on agar with a coloured dye. The plasmid that carries the human gene also has a second gene included in it. This gene codes for an enzyme that causes a colour change in a dye, turning it blue. One of two possible events will happen on the agar:

- If the human insulin gene is in the plasmid, it damages the gene that makes the colour-change enzyme. So no colour-change enzyme is produced, and the agar around the bacteria will stay colourless.
- If the human insulin gene is not present, the colour-change gene will be intact, and will make the enzyme. The result will be that the agar will go blue.

The colourless colonies of bacteria are selected and cloned.

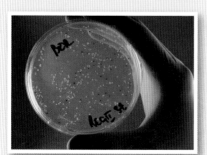

◀ An assaying technique to check whether the insulin gene has been taken up by bacteria. The blue colonies do not have the desired gene; the white colonies have taken up the gene and will be cultured.

Key words

biotechnology, genetic engineering, plasmid, transgenic

B What is a restriction enzyme?

C How does the DNA repair?

▲ Fermenter units containing bacteria that have been genetically engineered to produce proteins

Questions

1 What type of organism is the host for the insulin gene?

2 Why are genetically modified organisms called transgenic organisms?

3 Explain why biotechnology companies need to clone genetically modified bacteria.

4 Suggest how food, oxygen levels, and temperature could be managed inside a fermenter.

5 Why is it important to carry out an assay of the bacteria resulting from the genetic engineering process?

15: Genetically modified organisms

Key words

restriction enzymes, sticky end, DNA ligase

Improving on nature

Biotechnologists do not only put genes into bacteria. They can also transfer genes into other organisms, including plants and animals. These are then called genetically modified (GM) organisms. By adding new genes, their genetic code is altered. But why do this?

The idea is to take a gene coding for a useful characteristic from one organism and make it work in another (host), organism. The host will then show the useful characteristic.

Examples of characteristics transferred by GM technology include:

- pesticide resistance in plants
- frost resistance in fruit
- increased shelf life in fruit
- increased milk yield in cattle.

Creating a GM organism

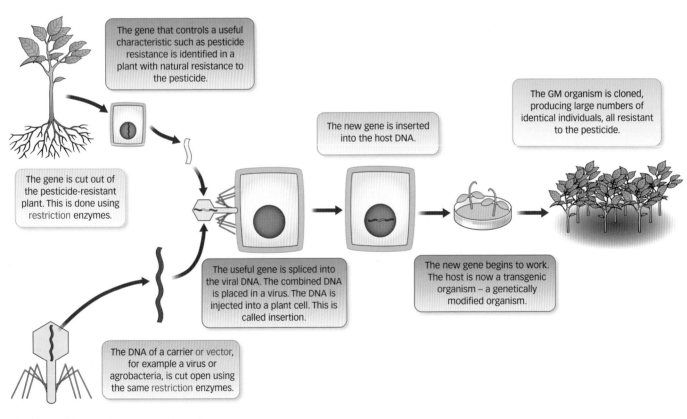

The gene that controls a useful characteristic such as pesticide resistance is identified in a plant with natural resistance to the pesticide.

The gene is cut out of the pesticide-resistant plant. This is done using restriction enzymes.

The DNA of a carrier or vector, for example a virus or agrobacteria, is cut open using the same restriction enzymes.

The useful gene is spliced into the viral DNA. The combined DNA is placed in a virus. The DNA is injected into a plant cell. This is called insertion.

The new gene is inserted into the host DNA.

The new gene begins to work. The host is now a transgenic organism – a genetically modified organism.

The GM organism is cloned, producing large numbers of identical individuals, all resistant to the pesticide.

▲ How GM organisms are produced

A What is a GM organism?

High-tech enzymes

It may seem surprising that biologists are able to put genes from one organism into another and they still work. The reason for this is that DNA and the genetic code is the same for all life. It is universal, and all cells can 'read' or 'understand' the same code. The trick is to use high-tech enzymes to ensure that the DNA for the genes is put into place correctly.

Restriction enzymes make very precise cuts through the DNA. Remember that DNA is made of two parallel strands, with bases between them. Restriction enzymes do not cut straight across the strands – most of them make a staggered cut, leaving a few unpaired bases exposed on the ends of the strands. The cut surfaces are called '**sticky ends**'.

The useful thing about sticky ends is that they will stick back together a little like Velcro. The new gene to be added is also cut by the same restriction enzyme, creating the same types of sticky ends. The new gene will therefore fit into the vector's cut DNA, and the sticky ends will come together. The enzyme **DNA ligase** repairs the paired DNA strands.

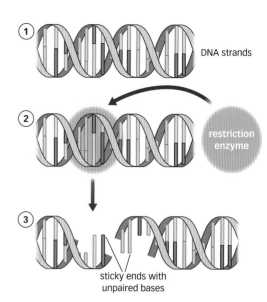

▲ Restriction enzymes cut DNA strands, leaving sticky ends

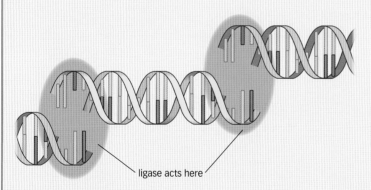

ligase acts here

▲ DNA ligase repairs DNA, incorporating the new gene

> **B** Name a common vector used to carry DNA into a plant cell.

Questions

1 Name some characteristics that might be transferred by GM technology.

2 What two things are enzymes used for during the process of genetic engineering?

3 Explain in detail how restriction enzymes cut DNA to create a sticky end.

4 Why are sticky ends useful during the process of genetic engineering?

5 What does DNA ligase do?

After studying this topic, you should be able to:

- ✔ know that everyone has a unique DNA sequence
- ✔ describe the process of DNA fingerprinting

◀ We each have unique fingerprints

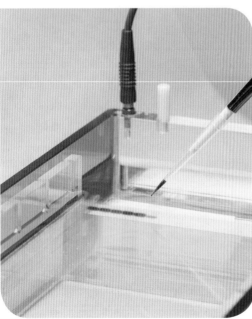

▲ Electrophoresis apparatus. The DNA fragments are placed in wells in a gel. The gel sits in a trough with a solution. An electric current passes through the solution. This causes the DNA fragments to move different amounts depending on their size.

Identifying suspects

Every individual has their own unique fingerprint. In the same way, we all have a unique sequence of DNA in all our cells. A **DNA fingerprint** is an image of certain parts of a person's DNA. It can be used to identify them. Forensic scientists use DNA samples collected at the scene of a crime to make a DNA fingerprint and identify the criminal.

> **A** What is the difference between a fingerprint and a DNA fingerprint?
>
> **B** Identical twins are clones. Would their DNA fingerprints be different? Explain why.

Making a DNA fingerprint

Forensic scientists prepare a DNA fingerprint using this process:

1. A sample of human tissue is collected from the scene of a crime, such as blood, skin, hair, or semen.
2. The DNA is extracted from the cells.
3. The DNA is cut into fragments using restriction enzymes.
4. Since each person has different DNA, the restriction enzymes cut in different places, producing fragments of different lengths.

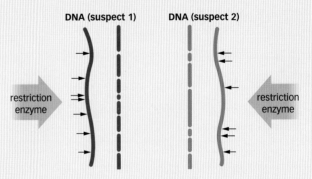

▲ A restriction enzyme recognises a particular sequence of bases

5. The fragments are separated using a kind of chromatography called electrophoresis. They are separated according to their size – smaller fragments travel further.

6. A series of colourless bands is produced on the gel.
7. To make the bands visible, a radioactive probe is added which sticks to the DNA. This can then be detected on film. The result is a film with a series of black and white bands. This is called an autoradiograph.
8. The forensic scientist compares the positions of the bands to identify the suspect.
9. The position of the bands should match if the suspect and the sample from the scene of the crime have the same DNA.

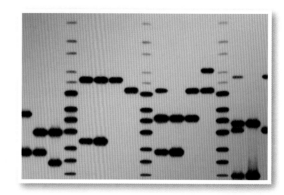

▲ An autoradiograph of DNA fragments. The radioactive probe bound to the different sized DNA fragments shows up on the film.

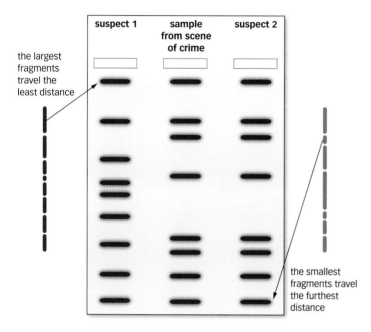

the largest fragments travel the least distance

the smallest fragments travel the furthest distance

▲ The DNA fingerprint from the scene of the crime sample is compared with the DNA fingerprints of suspects

Storing data

It is possible to store DNA fingerprints on computer records. This is useful for the police, as they can build up a large database of DNA fingerprints. It makes it easy to compare forensic evidence with the records of many potential suspects.

However, many people are concerned about storing this genetic information. They argue that storing the DNA fingerprints of innocent people is an invasion of privacy. They feel that their records should not be kept if they have not done anything wrong.

Key words

DNA fingerprint

Exam tip   OCR

✓ To interpret DNA fingerprints, look for similar band patterns between two samples.

Questions

1 Why can a DNA fingerprint be used to identify a person? ↓ E

2 Discuss the ethical issues about storing DNA fingerprint records. ↓ C

3 Explain why a radioactive gene probe is used in the process of DNA fingerprinting. ↓ A*

Module summary

Revision checklist

- Bacteria live in many different habitats. They reproduce by binary fission and can be grown in labs or fermenters.
- Fungi include yeast and mushrooms. Yeast reproduces by budding.
- Viruses do not have a cell structure. They have a protein coat around some genetic material. They are not truly living and have to use another organism's cell to make copies of themselves.
- Some microorganisms cause diseases. Skin, stomach acid, acid urine and the respiratory system help prevent infection.
- After natural disasters infections spread easily by means of dirty water and sewage.
- Bacteria cause cholera and food poisoning; viruses cause flu and chickenpox; fungi cause athlete's foot.
- Louis Pasteur, Joseph Lister, and Alexander Fleming developed ways of dealing with microbes.
- Bacteria are used to make yoghurt, cheese, vinegar, silage, and compost.
- Yeast is used to make wine and beer. Wine and beer can be distilled to make spirits (brandy and whisky).
- Plant material can be used for fuel (biofuels, wood, and biogas).
- Soil is made from particles of rock, water, humus, and air. Plants grow in soil. Earthworms live in soil.
- Many organisms live in water, which provides buoyancy and a stable temperature. Freshwater organisms need mechanisms to remove excess water.
- Plankton microorganisms form the basis of many food webs. On the ocean floor, bacteria form the basis of food webs.
- Human waste such as oil spills, fertilisers, sewage, and pesticides may pollute water and affect other organisms.
- Enzymes obtained from microorganisms are used to make processed foods, washing powders, cheese, and slimming products.
- Immobilised enzymes are used in some medical products.
- Genetically modified bacteria make useful medicinal products such as insulin.
- Plants and animals can be genetically modified.
- Forensic scientists can produce DNA fingerprints.

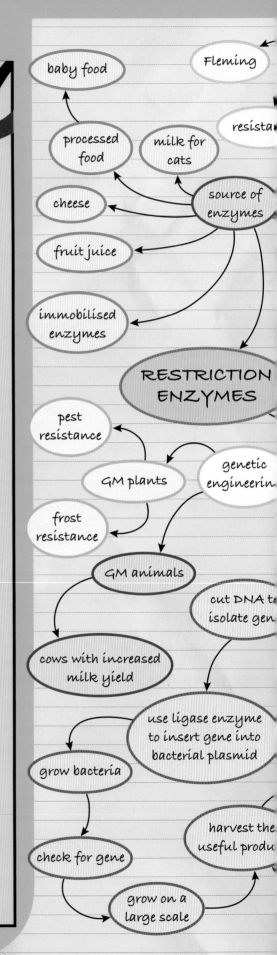

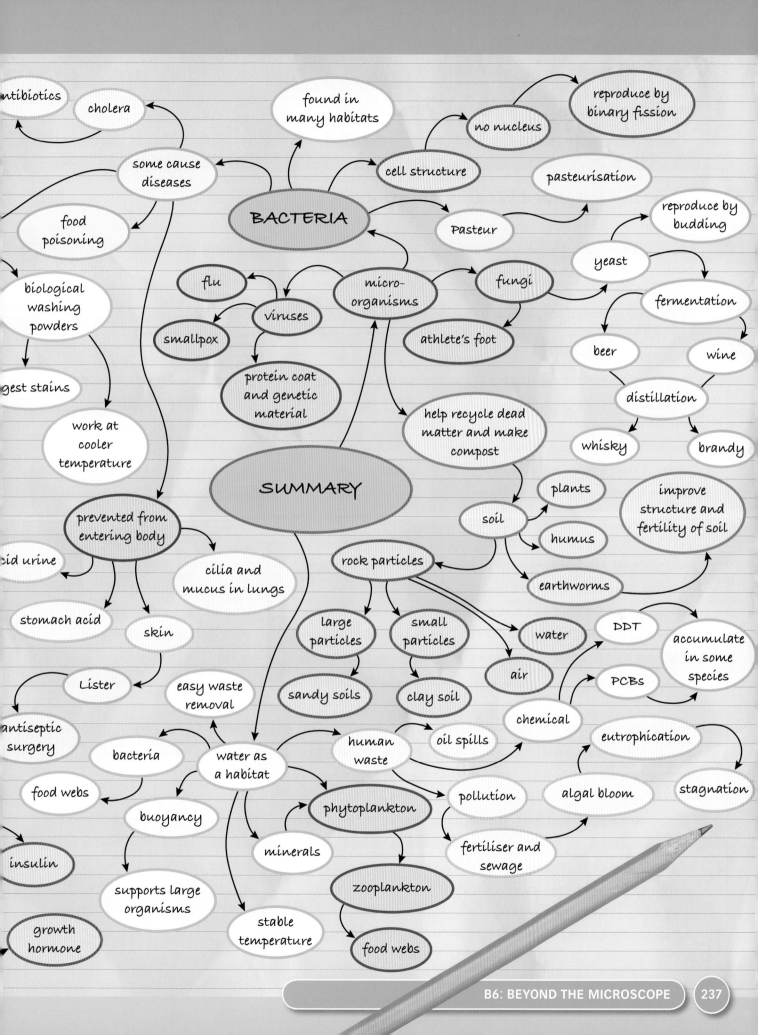

antibiotics

cholera

found in many habitats

no nucleus

reproduce by binary fission

some cause diseases

cell structure

pasteurisation

Pasteur

reproduce by budding

food poisoning

BACTERIA

yeast

biological washing powders

flu

micro-organisms

fungi

fermentation

viruses

beer

wine

digest stains

smallpox

athlete's foot

distillation

protein coat and genetic material

help recycle dead matter and make compost

whisky

brandy

work at cooler temperature

SUMMARY

plants

improve structure and fertility of soil

soil

humus

prevented from entering body

rock particles

earthworms

acid urine

cilia and mucus in lungs

DDT

accumulate in some species

stomach acid

skin

large particles

small particles

water

PCBs

Lister

easy waste removal

sandy soils

clay soil

air

antiseptic surgery

bacteria

water as a habitat

human waste

oil spills

chemical

eutrophication

food webs

buoyancy

phytoplankton

pollution

algal bloom

stagnation

insulin

minerals

zooplankton

fertiliser and sewage

supports large organisms

growth hormone

stable temperature

food webs

OCR gateway *Upgrade*

Answering Extended Writing questions

QUESTION

Describe the important contributions that scientists Pasteur, Lister, and Fleming made to the modern treatments of diseases. How can doctors and patients reduce the incidence of bacteria becoming resistant to antibiotics?

The quality of written communication will be assessed in your answer to this question.

I was away when we did Pasteur and Lister. Fleming invented penicillin.

People shouldnt take antibiotics unless they really need them or they will become resistant to them.

E

Examiner: Being away is no excuse for this candidate, as there is information in the student book! Fleming discovered penicillin, he didn't invent it. However, Fleming is linked to the correct discovery. The second part of this answer is confused. Bacteria become resistant to antibiotics, not people.

Pasteur said germs cause disease. Lister developed antiseptic surgery. Fleming discovered penicillin.

Doctors shouldnt give antibiotics for colds and flu as these are caused by viruses. Doctors should tell patients to take all their medicine and patients should take it all even if they feel better.

C

Examiner: The contribution of each scientist is briefly outlined. The second part is answered concisely. Very few grammatical errors.

People used to think diseases were caused by evil spirits or a punishment from god.

Pasteur studied silkworms and saw that microorganisms caused diseases. Doctors didn't believe him at first.

Lister used antiseptics during surgery to kill bacteria on skin. Before that hardly anyone survived surgery and lots of people died form infected wounds.

Fleming discovered penicillin. Bacteria wouldn't grow in a dish where there was a fungus.

Patients should take all their antibiotics even if they feel better.

A*

Examiner: A good introduction here sets the scene. The contributions of all three scientists are outlined clearly and correctly. The answer to the last part is brief but correct. The answer is well organised into paragraphs and there are no spelling or grammatical errors.

Exam-style questions

1 a i Enzymes are used in the food industry to make sugars sweeter. Name one other use of enzymes in the food industry.
A01

A01 **ii** Enzymes are present in diabetic reagent sticks. Why do people with diabetes use reagent sticks?
alginate lipase pesticide PCB

A01 **b** Yeast can be used to make wine. What process do the yeast cells carry out?
i photosynthesis
ii digestion
iii fermentation.

2 a *Salmonella* bacteria can cause food poisoning.

A01 **i** Describe how these bacteria enter the body.

A01 **ii** Food poisoning can be treated with antibiotics. What are antibiotics?

A01 **iii** Some bacteria are resistant to antibiotics. Describe how bacteria become resistant to antibiotics.

A01 **b** Explain why viruses are described as not really living.

A01 **c** List three ways in which the body tries to prevent infection by bacteria or viruses.

3 A student investigated the effect of temperature on yeast activity. The table shows her results.

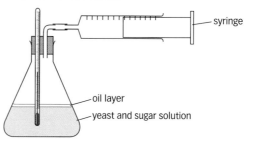
syringe
oil layer
yeast and sugar solution

temperature, °C	10	20	30	40	50	60
time to make 5 cm³ carbon dioxide, minutes	24	38	5	2	15	56

A02 **a** Which result is anomalous?

A02 **b** Why was the layer of oil placed over the yeast and sugar solution?

A02 **c** How did she make the investigation valid (fair)?

Extended Writing

4 Describe what soil is made of. Why are earthworms important in the soil?
A01

5 Describe the composition of biogas and how it is made.
A01

6 State two advantages and two disadvantages of using biofuels.
A01

7 Describe how DNA fingerprinting is carried out. Why are some people concerned about a national database to store people's DNA?
A01

A01 Recall the science
A02 Apply your knowledge
A03 Evaluate and analyse the evidence

Revising module B1

To help you start your revision, the specification for module B1 has been summarised in the checklist below. Work your way along each row and make sure that you are happy with all the statements for your target grade.

If you are not sure of any of the statements for your target grade, make a note of them as part of your revision plan. You can then work back through the relevant parts of pages 14–45 to fill gaps in your knowledge as a start to your revision.

To aim for a grade G–E	To aim for a grade D–C	To aim for a grade B–A*
B1a **Explain** why blood in arteries is under pressure. **Recognise** that the risk of developing heart disease can be increased by a number of factors. **Describe** how cholesterol can restrict or block arteries.	**Recall** what blood pressure measurements consist of. **Describe** the factors that increase and decrease blood pressure. **Explain** the difference between fitness and health. **Analyse** the results of different ways of measuring fitness. **Explain** how smoking increases blood pressure. **Explain** how diet can reduce the risk of heart disease.	**Explain** the consequences of high blood pressure. **Explain** the consequences of low blood pressure. **Evaluate** different ways of measuring fitness. **Explain** why carbon dioxide reduces the carrying capacity of red blood cells. **Explain** how narrowed coronary arteries and thrombosis increase the risk of a heart attack.
B1b **Explain** what a balanced diet should contain and why. **Interpret** simple data on diet. **Explain** why teenagers need a high protein diet. **Explain** why diets in many parts of the world are deficient in protein. **Recall** that proteins are only used as an energy source in a shortage. **Recall** that obesity is linked to increased health risks.	**Recall** what carbohydrates, fats, and proteins are made of. **Explain** the factors involved in how a balanced diet will vary from person to person. **Explain** why protein deficiency is common in developing countries. **Calculate** EAR for protein. **Calculate** and **understand** BMI measurements. **Explain** how poor self-image can lead to a poor diet.	**Describe** the storage of carbohydrates, fats, and proteins. **Describe** the difference between first and second class proteins. **Understand** that EAR is an estimated figure for an average person. **Explain** why EAR for protein may vary from person to person.
B1c **Recall** that infectious diseases are caused by pathogens. **Recall** examples of diseases caused by different types of pathogen. **Describe** the human body's defences against pathogens. **Describe** the difference between infectious and noninfectious diseases. **Understand** that some disorders have other causes, including genetic causes. **Recall** that immunisation gives protection against certain pathogens. **Describe** how pathogens are destroyed by the body's immune system. **Interpret** data on the incidence of disease. **Explain** why new treatments are tested before use.	**Recall** the meaning of the terms parasite and host with reference to malaria. **Describe** how vectors carry disease. **Describe** changes in lifestyle and diet that may reduce the risk of cancer. **Explain** how pathogens cause the symptoms of a disease. **Recall** how antibodies fight pathogens. **Explain** the difference between passive and active immunity. **Recall** the difference between antibiotics and antiviral drugs. **Describe** how new treatments are tested. **Understand** objections to some forms of testing.	**Explain** how knowledge of vectors can help control infections. **Interpret** data on types of cancer. **Describe** the difference between benign and malignant tumours. **Explain** why each pathogen needs specific antibodies. **Explain** the process of immunisation. **Describe** the benefits and risks of immunisation. **Explain** the need to prevent the spread of antibiotic resistance. **Explain** why blind and double blind trials are used in testing new treatments.
B1d **Describe** how animals detect stimuli using receptors. **Name** and **locate** the main parts of the eye. **Explain** the advantages and disadvantages of monocular and binocular vision. **Describe** the most common vision problems. **Name** and **locate** the main parts of the nervous system. **Describe** nerve impulses. **Describe** reflex actions. **Recognise** that voluntary responses are under the conscious control of the brain.	**Describe** the functions of the main parts of the eye. **Describe** the pathway of light through the eyeball. **Explain** how binocular vision helps to judge distances. **Explain** the cause of common vision problems. **Name** and **locate** the parts of a motor neurone. **Recall** that the nerve impulse passes along the axon. **Describe** a reflex arc. **Describe** the path of a spinal reflex.	**Explain** how the eye focuses light from distant and close objects. **Explain** how long and short sight can be corrected. **Explain** how neurones are adapted to their function. **Recall** that the gap between neurones is called a synapse. **Describe** transmission of a nerve impulse across a synapse.

To aim for a grade G–E	To aim for a grade D–C	To aim for a grade B–A*	
Recognise that some drugs can be beneficial or harmful. **Explain** why some drugs are only available on prescription. **Explain** addiction, withdrawal symptoms, tolerance, and rehabilitation. **Describe** the general effects of the different drug categories. **Recall** the health problems caused by smoking. **Describe** the effects of carbon monoxide, nicotine, tars, and particulates on the body. **Recognise** short and long term effects of alcohol on the body. **Explain** why there is a legal alcohol limit.	**Explain** the basis for the legal classification of drugs. **Recall** examples of the different drug categories. **Describe** the effect of cigarette smoke on ciliated epithelial cells. **Explain** why damage to ciliated epithelial cells can cause 'smoker's cough'. **Interpret** data on the alcohol content of different drinks. **Interpret** information on reaction times, accident statistics, and alcohol levels.	**Explain** the action of depressants and stimulants on the synapses of the nervous system. **Evaluate** data on the effects of smoking. **Describe** how alcohol can cause cirrhosis of the liver.	B1e
Recognise that the body works to maintain a steady state. **Recall** that the normal core human body temperature is 37°C. **Describe** how to measure body temperature. **Describe** how heat can be gained or lost. **Name**, **locate**, and **recall** the function of the pancreas. **Recall** the cause of Type 1 diabetes. **Describe** how insulin travels around the body.	**Understand** homeostasis and what it involves. **Explain** why internal conditions are kept steady by automatic control systems. **Explain** how sweating increases heat transfer to the environment. **Understand** that body temperature is the optimum temperature for many enzymes. **Describe** the problems caused by extreme body temperatures. **Recall** that insulin controls blood sugar levels. **Explain** how Type 1 and Type 2 diabetes are controlled. **Explain** why responses controlled by hormones are usually slower than those controlled by the nervous system.	**Explain** how negative feedback mechanisms maintain a constant internal environment. **Explain** how vasodilation and vasoconstriction work. **Understand** that optimum body temperature is linked to enzyme action. **Explain** that the brain monitors blood temperature and triggers temperature control mechanisms. **Explain** how insulin regulates blood sugar levels. **Explain** how insulin dosage varies between Type 1 diabetes patients.	B1f
Recognise that plants and animals respond to environmental changes. **Understand** that plant growth is controlled by plant hormones. **Describe** an experiment showing that shoots grow towards light. **Understand** how growth towards light increases the plant's chance of survival. **Understand** that roots grow downwards due to gravity. **Recognise** that plant hormones are used in agriculture to control plant growth.	**Describe** shoots and roots in terms of phototropism and geotropism. **Recall** the action of auxins. **Relate** the action of plant hormones to their commercial uses.	**Interpret** data from phototropism experiments. **Explain** how auxin causes shoot curvature in terms of cell elongation.	B1g
Analyze human characteristics to **determine** those that are a result of both environmental and inherited factors. **Recall** that chromosomes in the nucleus carry genes controlling inherited characteristics. **Recognise** that most body cells contain matching pairs of chromosomes. **Recall** that gametes have half the number of chromosomes of body cells. **Recognise** that some disorders are inherited.	**Identify** inherited characteristics as dominant or recessive. **Explain** the causes of genetic variation. **Recall** that the number of chromosomes in body cells varies between species. **Recall** that alleles are different versions of the same gene. **Describe** how sex in mammals is determined by the XX and XY sex chromosomes. **Understand** that inherited disorders are caused by faulty genes. **Understand** the issues raised by knowledge of inherited disorders in a family.	**Understand** the debate over the relative importance of genetic and environmental factors. **Explain** the link between dominant and recessive characteristics and alleles. **Explain** a monohybrid cross. **Use** and **explain** the terms homozygous, heterozygous, genotype, and phenotype. **Explain** sex inheritance using genetic diagrams. **Recall** that inherited disorders are caused by faulty alleles. **Use** genetic diagrams to **make predictions** about inherited disorders.	B1h

Revising module B2

To help you start your revision, the specification for module B2 has been summarised in the checklist below. Work your way along each row and make sure that you are happy with all the statements for your target grade.

If you are not sure of any of the statements for your target grade, make a note of them as part of your revision plan. You can then work back through the relevant parts of pages 52–83 to fill gaps in your knowledge as a start to your revision.

To aim for a grade G–E	To aim for a grade D–C	To aim for a grade B–A*
B2a **Understand** classification of organisms into groups. **Describe** how organisms are placed into the five Kingdoms. **Use** characteristics to place organisms into the different classes of arthropods. **Recognise** that organisms within a species may show great variation. **Understand** why similar species often share similar habitats.	**Understand** the difficulties in placing organisms into distinct groups. **Describe** classification of organisms. **Explain** the importance of classifying species. **Define** the term species. **Understand** how evolutionary relationships between organisms can be displayed. **Explain** the importance of the binomial system for naming species. **Recall** that closely related species share a common ancestor and may have different features.	**Describe** natural and artificial classification systems. **Explain** how genetic information has changed understanding of classification. **Understand** why classification systems change over time. **Understand** how evolutionary relationships can be modelled. **Explain** some of the problems of classifying organisms into species. **Explain** how similarities and differences can be explained in terms of evolution and ecology.
B2b **Explain** the term trophic level. **Understand** that organisms other than green plants are producers. **Explain** why some organisms are both primary and secondary consumers. **Explain** how population change may have an effect throughout a food web. **Explain** how energy from the Sun flows through food webs. **Interpret** data on energy flow in food webs.	**Understand** what pyramids of biomass show. **Construct** pyramids of biomass. **Explain** the differing shapes of pyramids of numbers and pyramids of biomass for the same food chain. **Explain** how energy is lost at each stage of the food chain. **Describe** how excretory products and uneaten parts can form the start of new food chains.	**Explain** the difficulties in constructing pyramids. **Explain** how efficiency of energy transfer explains pyramid shape. **Explain** how efficiency of energy transfer limits the length of food chains. **Calculate** the efficiency of energy transfer.
B2c **Recall** that the elements in plants and animals are recycled through decay. **Recognise** that many soil bacteria and fungi are decomposers. **Describe** the importance of the decay process. **Recognise** that animals and plants take in elements from chemicals as they grow. **Recall** that carbon and nitrogen are two of the most important elements. **Recall** that plants take up carbon as carbon dioxide. **Recall** that nitrogen is taken up by plants as nitrates. **Recall** the abundance of nitrogen in the air. **Explain** why nitrogen gas can't be used directly by animals or plants.	**Explain** why nutrient recycling takes longer in waterlogged or acidic soils. **Explain** how carbon is recycled in nature. **Explain** how nitrogen is recycled in nature.	**Explain** how carbon is recycled in nature, to include carbon sinks. **Explain** how nitrogen is recycled in nature, to include the conversion of ammonia to nitrates.

To aim for a grade G–E	To aim for a grade D–C	To aim for a grade B–A*	
Explain how competition influences animal distribution and population size. **Interpret** data on competition for resources. **Explain** the relationship between the size of predator and prey populations. **Recall** the beneficial relationships between some organisms of different species and **describe** an example.	**Explain** close competition between similar animals in the same habitat. **Describe** competition between organisms within a species. **Explain** how some predator-prey populations show cyclical fluctuations. **Describe** parasitism and mutualism.	**Describe** competition as interspecific or intraspecific. **Explain** why intraspecific competition is often more significant. **Explain** the term ecological niche, and **understand** that similar organisms will occupy similar niches. **Explain** why predator and prey population cycles are out of phase. **Explain** how interdependence of organisms determines their distribution and abundance. **Explain** why nitrogen-fixing bacteria in root nodules are an example of mutualism.	**B2d**
Explain how predators are adapted for success. **Explain** how prey animals are adapted to evade being caught. **Recall** that better-adapted organisms are better able to compete.	**Explain** some adaptations to cold environments. **Explain** some adaptations to hot environments. **Explain** some adaptations to dry environments. **Explain** how better-adapted organisms are better able to compete.	**Explain** how counter-current heat exchange systems work. **Understand** that some organisms are biochemically adapted to extreme conditions. **Analyse** surface area to volume ratios in the context of environmental stresses. **Explain** how adaptations help organisms cope with environmental stresses. **Describe** how some organisms are specialists and some are generalists.	**B2e**
Identify variation within a population of organisms of one species. **Explain** why better adapted animals and plants are more likely to survive. **Recognise** evolution as change in groups of organisms over time. **Understand** how some organisms survive environmental change while others become extinct. **Recall** that whilst many theories of evolution have been put forward, most accept that of Darwin.	**Understand** Darwin's theory of evolution by natural selection. **Recall** that adaptations are controlled by genes that can be inherited. **Explain** why the theory of natural selection was initially rejected. **Recognise** that natural selection is now generally accepted.	**Explain** how the changes brought about by natural selection may result in new species. **Understand** why speciation requires geographical or reproductive isolation. **Explain** the differences between the theories of Darwin and Lamarck. **Explain** why Lamarck's theory was discredited. **Recognise** that the theory of natural selection has developed in line with new discoveries.	**B2f**
Recognise human population increase. **Recognise** that the human population uses resources. **Explain** that an increasing population leads to increasing pollution and use of resources. **Understand** that pollution can affect the number and type of organisms in a particular place.	**Understand** that the human population is increasing exponentially. **Understand** that population growth is the result of increasing birth and decreasing death rates. **Explain** the causes and consequences of global warming, ozone depletion, and acid rain. **Explain** how indicator species help indicate levels of pollution. **Describe** how pollution can be measured.	**Explain** how developed countries use the most resources and cause the most pollution. **Explain** the term carbon footprint. **Discuss** the possible consequences of exponential growth. **Interpret** data on indicator species. **Describe** the advantages and disadvantages of living and non-living methods of measuring pollution.	**B2g**
Explain why organisms become extinct or endangered. **Describe** how endangered species can be helped. **Interpret** data on whale species' distributions. **Discuss** why certain whale species are close to extinction. **Recognise** what a sustainable resource is. **Recall** that some resources can be maintained.	**Explain** reasons for conservation programmes. **Explain** why species are at risk of extinction if numbers fall below a critical level. **Recognise** the value of both living and dead whales. **Describe** issues arising from keeping whales in captivity. **Explain** the term sustainable development. **Explain** how fish stocks and woodland can be sustained and developed.	**Explain** why species are at risk of extinction if there is not enough genetic variation. **Evaluate** a given example of a conservation programme. **Recognise** that some aspects of whale biology are not fully understood. **Describe** issues concerning whaling. **Explain** the importance of population size, waste products, food, and energy in sustainable development. **Understand** that sustainability requires planning and co-operation. **Describe** how sustainable development may protect endangered species.	**B2h**

Revising module B3

To help you start your revision, the specification for module B3 has been summarised in the checklist below. Work your way along each row and make sure that you are happy with all the statements for your target grade.

If you are not sure of any of the statements for your target grade, make a note of them as part of your revision plan. You can then work back through the relevant parts of pages 90–121 to fill gaps in your knowledge as a start to your revision.

To aim for a grade G–E	To aim for a grade D–C	To aim for a grade B–A*
B3a **Identify** the mitochondria in an animal cell. **Recall** that respiration occurs in the mitochondria. **Recall** that chromosomes in the nucleus carry coded information in DNA. **Recall** that the information in genes is called the genetic code. **Understand** that the genetic code controls cell activity and some characteristics. **Recall** that DNA controls the production of different proteins. **Recall** that proteins are needed for the growth and repair of cells. **Recall** that the structure of DNA was first worked out by Watson and Crick.	**Explain** why liver and muscle cells have large numbers of mitochondria. **Describe** the structure of DNA as a double helix with cross links formed by pairs of bases. **Describe** chromosomes as long, coiled molecules of DNA, divided up into genes. **Recall** that each gene contains a different sequence of bases. **Recall** that each gene codes for a protein. **Recall** that proteins are made in the cytoplasm. **Understand** why a copy of the gene is needed. **Describe** how Watson and Crick used data from other scientists to build a model of DNA.	**Recall** that some structures in cells are too small to be seen with a light microscope. **Recall** that ribosomes are in the cytoplasm and are the site of protein synthesis. **Recall** the four bases of DNA. **Describe** complementary base pairings. **Explain** how protein structure is determined. **Explain** how the code needed to produce a protein is carried from the DNA to the ribosomes by a molecule called mRNA. **Explain** how DNA controls cell function by controlling the production of proteins. **Explain** why new discoveries are not accepted immediately, to include the importance of repeating or testing the work.
B3b **Recall** some examples of proteins, to include collagen, insulin, and haemoglobin. **Describe** enzymes as proteins. **Understand** that enzymes have active sites that substrate molecules fit into when a reaction takes place. **Recognise** that different cells and different organisms will produce different proteins. **Describe** gene mutations as changes to genes.	**Recognise** that proteins are made of long chains of amino acids. **Describe** some functions of proteins. **Describe** enzymes as biological catalysts. **Explain** the specificity of enzymes. **Describe** how changing temperature and pH will affect an enzyme-catalysed reaction. **Recall** that mutations may lead to the production of different proteins. **Understand** how mutations occur. **Understand** that mutations are often harmful but may be beneficial.	**Explain** how each protein has its own number and sequence of amino acids. **Explain** how enzyme activity is affected by pH and temperature. **Calculate** and **interpret** the Q10 value for a reaction over a 10°C interval. **Understand** that only some of the full set of genes are used in any one cell. **Understand** that the genes switched on determine the functions of the cell. **Explain** how changes to genes alter or prevent the production of proteins.
B3c **Recognise** that the energy provided by respiration is needed for all life processes in plants and animals. **Recall** and **use** the word equation for aerobic respiration. **Describe** examples of life processes that require energy from respiration. **Explain** why breathing and pulse rates increase during exercise. **Describe** an experiment to measure resting pulse rate and recovery time after exercise. **Analyse** given data from a pulse rate experiment.	**Recall** and **use** the symbol equation for aerobic respiration. **Use** data from experiments to compare respiration rates. **Calculate** the respiratory quotient. **Explain** why anaerobic respiration takes place during hard exercise. **Recall** that anaerobic respiration produces lactic acid. **Recall** and **use** the word equation for anaerobic respiration. **Understand** that anaerobic respiration releases less energy per molecule of glucose.	**Recall** that respiration results in the production of ATP, used as the energy source for many processes in cells. **Explain** how the rate of oxygen consumption can be used as an estimate of metabolic rate. **Explain** why the rate of respiration is influenced by changes in temperature and pH. **Explain** fatigue in terms of lactic acid build up and how this is removed during recovery.
B3d **Describe** the difference between unicellular and multicellular organisms. **Recall** that most body cells contain chromosomes in matching pairs.	**Explain** the advantages of being multicellular. **Recall** that new cells for growth are produced by mitosis.	**Explain** why becoming multicellular requires the development of specialised organ systems.

To aim for a grade G–E	To aim for a grade D–C	To aim for a grade B–A*	
Explain why the chromosomes have to be copied to produce new cells for growth. **Recall** that this type of cell division is also needed to replace cells, repair tissues and for asexual reproduction. **Recall** that in sexual reproduction gametes join in fertilisation. **Recall** that gametes have half the number of chromosomes of body cells. **Understand** that in sexual reproduction half the genes come from each parent. **Explain** why many sperm cells are produced.	**Explain** why these new cells are genetically identical. **Recall** that in mammals, body cells are diploid. **Explain** why DNA replication must take place before cells divide. **Recall** that gametes are produced by meiosis. **Describe** gametes as haploid. **Explain** why fertilisation results in genetic variation. **Explain** the structure of a sperm cell.	**Describe** how DNA replication occurs prior to mitosis. **Describe** how the chromosomes behave in mitosis. **Explain** why the chromosome number is halved in meiosis and each cell is genetically different.	B3d
Describe the functions of different blood cells. **Describe** how to make a slide of an onion cell. **Recall** that the blood moves around the body in arteries, veins, and capillaries. **Recall** that the blood moves around the body in arteries, veins and capillaries. **Describe** the functions of the right side and the left side of the heart in the pumping of blood. **Recall** that blood in arteries is under higher pressure than the blood in the veins. **Explain** why blood flows from one area to another in terms of pressure difference.	**Explain** how the structure of a red blood cell is adapted to its function. **Describe** the function of plasma. **Describe** how the parts of the circulatory system work together to transport substances. **Identify** the names and positions of the parts of the heart and describe their functions. **Explain** why the left ventricle has a thicker muscle wall than the right ventricle.	**Explain** the structure of a red blood cell in terms of surface area and volume. **Describe** how haemoglobin in red blood cells reacts with oxygen in the lungs forming oxyhaemoglobin, and that the reverse of this reaction happens in the tissues. **Explain** how the adaptations of arteries, veins, and capillaries relate to their functions. **Explain** the advantages of the double circulatory system in mammals.	B3e
Describe the functions of parts of a plant cell. **Understand** that bacterial cells are smaller and simpler than plant and animal cells. **Recall** that growth can be measured as an increase in height, wet mass, or dry mass. **Interpret** data ona typical growth curve. **Describe** the process of growth as cell division followed by cells becoming specialised. **Recall** that the process of cells becoming specialised is called differentiation. **Understand** that animals grow in the early stages of their lives and plants grow continually. **Understand** that plants grow at specific parts.	**Identify** simple differences between bacterial and plant and animal cells. **Recall** that bacterial cells lack a 'true' nucleus, mitochondria, and chloroplasts. **Recall** that dry mass is the best measure of growth. **Describe** a typical growth curve. **Recall** that in human growth there are two phases of rapid growth. **Recall** that stem cells differentiate. **Recall** that stem cells can be obtained from embryonic tissue and could potentially be used to treat medical conditions. **Discuss** issues arising from stem cell research in animals. **Explain** why plant and animal growth differs.	**Describe** the difference between the arrangement of DNA in a bacterial cell and a plant or animal cell. **Explain** the advantages and disadvantages of measuring growth by length, wet mass, and dry mass. **Explain** why the growth of different parts of an organism may differ from the growth rate of the whole organism. **Explain** the difference between adult and embryonic stem cells.	B3f
Describe the process of selective breeding. **Explain** how selective breeding can contribute to improved agricultural yields. **Recall** that selected genes can be artificially transferred by genetic engineering, producing organisms with different characteristics. **Identify** features that might be selected for in a genetic engineering programme. **Recognise** that in the future it may be possible to change a person's genes and cure disorders.	**Recognise** that a selective breeding programme may lead to inbreeding that can cause health problems within the species. **Explain** some potential advantages and risks of genetic engineering. **Describe**, in outline only, some examples of genetic engineering. **Discuss** the ethical issues involved in genetic modification. **Recall** that changing a person's genes to cure disorders is called gene therapy.	**Explain** how a selective breeding programme may reduce the gene pool. **Understand** the principles of genetic engineering. **Recall** that gene therapy could involve body cells or gametes. **Explain** why gene therapy involving gametes is controversial.	B3g
Recall that cloning is asexual reproduction producing genetically identical copies. **Recall** that Dolly the sheep was the first mammal cloned from an adult. **Recognise** that identical twins are naturally occurring clones. **Recognise** that plants grown from cuttings or tissue culture are clones. **Describe** how some plants reproduce asexually. **Describe** how to take a cutting.	**Understand** the process of nuclear transfer that was used to produce Dolly. **Describe** some possible uses of cloning. **Understand** the ethical dilemmas concerning human cloning. **Describe** the advantages and disadvantages of the commercial use of cloned plants.	**Describe** the cloning technique used to produce Dolly. **Describe** the benefits and risks of cloning. **Explain** the possible implications of using animals to supply replacement organs. **Discuss** the ethical dilemmas concerning human cloning. **Describe** plant cloning by tissue culture. **Explain** why cloning plants is easier than cloning animals.	B3h

Revising module B4

To help you start your revision, the specification for module B4 has been summarised in the checklist below. Work your way along each row and make sure that you are happy with all the statements for your target grade.

If you are not sure of any of the statements for your target grade, make a note of them as part of your revision plan. You can then work back through the relevant parts of pages 128–159 to fill gaps in your knowledge as a start to your revision.

To aim for a grade G–E	To aim for a grade D–C	To aim for a grade B–A*
B4a **Describe** how to use collecting/counting methods, to include pooters, nets, pitfall traps, and quadrats. **Describe** a method to show that a variety of plants and animals live in a small area. **Use** keys to identify plants and animals. **Explain** how the distribution of organisms within a habitat can be affected. **Define** biodiversity as the variety of different species living in a habitat. **Identify** natural ecosystems and artificial ecosystems.	**Use** data from collecting/counting methods to calculate an estimate of population size. **Explain** the differences between ecosystem and habitat, community and population. **Describe** how to map the distribution of organisms in a habitat using a transect line. **Interpret** distribution data from kite diagrams **Compare** the biodiversity of natural and artificial ecosystems.	**Explain** the effect of sample size on accuracy. **Explain** the need to make certain assumptions when using capture-recapture data. **Explain** what it means for an ecosystem to be described as self-supporting. **Describe** zonation as a gradual change in the distribution of species across a habitat. **Explain** how a gradual change in an abiotic factor can result in the zonation of organisms. **Explain** reasons for the differences between the biodiversity of different habitats.
B4b **Recall** and use the word equation for photosynthesis. **Understand** that oxygen is a waste product in this reaction. **Recall** how glucose made in photosynthesis is transported and stored. **Recall** that glucose and starch can be converted to other substances. **Explain** why plants grow faster in summer. **Understand** that plants carry out respiration as well as photosynthesis.	**Recall** and use the balanced symbol equation for photosynthesis. **Describe** the development of the understanding of the process of photosynthesis. **Describe** the conversion of glucose and starch to other substances in plants, and their use. **Describe** how photosynthesis can be increased by changing carbon dioxide, light, and temperature levels. **Explain** why plants carry out respiration all the time.	**Explain** how experiments using isotopes have increased our understanding of photosynthesis. **Describe** photosynthesis as a two stage process. **Explain** why insoluble substances such as starch are used for storage. **Explain** the effects of limiting factors on the rate of photosynthesis. **Explain** why plants take in carbon dioxide and give out oxygen during the day and do the reverse at night.
B4c **Understand** why chloroplasts are not found in all plant cells. **Recall** that chlorophyll pigments in chloroplasts absorb light energy. **Recall** the entry points of materials required for photosynthesis. **Recall** the exit point of materials produced in photosynthesis. **Understand** that broader leaves enable more sunlight to be absorbed.	**Name** and **locate** the parts of a leaf. **Explain** how leaves are adapted for efficient photosynthesis.	**Explain** how the cellular structure of a leaf is adapted for efficient photosynthesis. **Interpret** data on the absorption of light to **explain** how plants maximise the use of energy from the Sun.
B4d **Recall** that substances move in and out of cells by diffusion and **describe** diffusion. **Recognise** that water moves in and out of plant cells by osmosis. **Recall** that the plant cell wall provides support. **Understand** that lack of water can cause plants to droop. **Describe** how carbon dioxide and oxygen diffuse in and out of plants. **Understand** that water moves in and out of animal cells through the cell membrane.	**Explain** the net movement of particles by diffusion. **Describe** diffusion through the cell membrane. **Describe** osmosis. **Recall** that osmosis is a type of diffusion. **Explain** the term partially permeable. **Explain** how plants are supported. **Explain** wilting. **Explain** how leaves are adapted to increase the rate of diffusion of carbon dioxide and oxygen. **Describe** the effects of the uptake and loss of water on animal cells.	**Explain** how the rate of diffusion is increased. **Explain** the net movement of water molecules by osmosis from an area of high water concentration to an area of low water concentration. **Predict** the direction of water movement in osmosis. **Explain** the terms flaccid, plasmolysed, and turgid. **Explain** why there are differences in the effects of water uptake and loss on plant and animal cells. **Use** the terms crenation and lysis.

To aim for a grade G–E	To aim for a grade D–C	To aim for a grade B–A*	
Relate plant structure to function. **Describe** how water travels through a plant. **Describe** experiments to show how transpiration rate can be affected. **Understand** that healthy plants must balance water loss with water uptake.	**Describe** the arrangement of xylem and phloem in a dicotyledonous root, stem, and leaf. **Relate** xylem and phloem to their function. **Understand** that both xylem and phloem form continuous systems in leaves, stems, and roots. **Recall** that transpiration is the evaporation and diffusion of water from inside leaves. **Describe** how transpiration causes water to be moved up xylem vessels. **Describe** factors affecting transpiration rate. **Explain** how root hairs increase the ability of roots to take up water by osmosis. **Recall** that transpiration provides plants with water for cooling, photosynthesis, support, and movement of minerals. **Explain** how the structure of a leaf is adapted to reduce excessive water loss.	**Describe** the structure of xylem and phloem. **Explain** how transpiration and water loss from leaves are a consequence of the way in which leaves are adapted for efficient photosynthesis. **Explain** why transpiration rate is increased by changes in light, temperature, air movement, and humidity. **Explain** how the cellular structure of a leaf is adapted to reduce water loss.	B4e
Recall that fertilisers contain minerals and that these are needed for plant growth. **Interpret** data on NPK values. **Describe** experiments to show the effects of mineral deficiencies on plants. **Describe** how minerals (including those dissolved in solution) are absorbed by the root hairs and from the soil.	**Explain** why plants require certain minerals for growth and photosynthesis. **Relate** mineral deficiencies to the poor plant growth that results. **Recall** that minerals are usually present in soil in quite low concentrations.	**Describe** how elements obtained from soil minerals are used in the production of compounds in plants. **Explain** how minerals are taken up into root hair cells by active transport. **Understand** that active transport can move substances from low concentrations to high concentrations using energy from respiration.	B4f
Recall the key factors in the process of decay. **Explain** why decay is important for plant growth. **Describe** how to carry out an experiment to show that decay is caused by the decomposers bacteria and fungi. **Recall** that microorganisms are used to break down human and plant waste. **Recognise** that food preservation techniques reduce the rate of decay.	**Describe** the effects of changing temperature, amount of oxygen, and amount of water on rate of decay. **Recall** that detritivores including earthworms, maggots and woodlice feed on dead and decaying material. **Explain** how food preservation methods reduce the rate of decay.	**Explain** why changing temperature and the amounts of oxygen and water affect the rate of decay. **Explain** the term saprophyte. **Explain** how saprophytic fungi digest dead material in terms of extracellular digestion.	B4g
Analyse data to show that farmers can produce more food if they use pesticides but that these can harm the environment. **Recall** that pesticides kill pests (any organisms that damage crops). **Recall** that examples of pesticides include insecticides, fungicides, and herbicides. **Recall** that intensive farming aims to produce as much food as possible. **Describe** how intensive farming methods can increase productivity. **Describe** organic farming methods. **Describe** how pests can be controlled biologically by introducing predators.	**Explain** the disadvantages of using pesticides. **Describe** how plants can be grown without soil (hydroponics). **Describe** possible uses of hydroponics. **Understand** that intensive farming methods may be efficient, but they raise ethical dilemmas. **Describe** organic farming techniques. **Explain** the advantages and disadvantages of biological control and organic farming. **Explain** how removing one or more organisms from a food chain or web may affect other organisms.	**Explain** the advantages and disadvantages of hydroponics. **Explain** how intensive food production improves the efficiency of energy transfer.	B4h

Revising module B5

To help you start your revision, the specification for module B5 has been summarised in the checklist below. Work your way along each row and make sure that you are happy with all the statements for your target grade.

If you are not sure of any of the statements for your target grade, make a note of them as part of your revision plan. You can then work back through the relevant parts of pages 166–197 to fill gaps in your knowledge as a start to your revision.

To aim for a grade G–E	To aim for a grade D–C	To aim for a grade B–A*
B5a **Recall** that different animals have different types of skeletons. **Recall** that an insect's external skeleton is made of chitin. **Describe** different internal skeleton types. **Describe** the different types of fractures of bones. **Recall** that X-rays detect fractures. **Describe** a joint as the place where two or more bones meet. **Recognise** that muscles move bones. **Identify** joints in the human body. **Identify** the main arm bones and muscles.	**Explain** why an internal skeleton is advantageous. **Understand** that cartilage and bone are living. **Describe** the structure of a long bone. **Explain** why hollow long bones are advantageous. **Understand** that, despite being very strong, bones can easily be broken by a sharp knock. **Explain** why the elderly are more to fractures. **Describe** the structure of synovial joints. **Describe** the types and range of movement in a ball and socket joint and a hinge joint. **Describe** how the biceps and triceps operate.	**Understand** that cartilage and bone are susceptible to infection, but can grow and repair themselves. **Describe** how, in humans, the skeleton starts off as cartilage but is later ossified. **Explain** why it can be dangerous to move a person with a suspected fracture **Explain** the functions in a synovial joint of synovial fluid and membrane, cartilage, and ligaments. **Explain** how the arm is an example of a lever.
B5b **Recall** details about different animals' circulatory systems. **Understand** the difference between open and closed circulatory systems. **Recall** that in a closed circulatory system blood will flow in arteries, veins, and capillaries. **Understand** how the heart muscle causes blood to move. **Describe** the heart as being made of powerful muscles that are supplied with food substances. **Understand** why the heart needs a constant supply of glucose and oxygen. **Describe** the pulse as a measure of heart beat and **recognise** that it can be detected at various places.	**Explain** why many animals need a blood circulatory system. **Describe** a single circulatory system. **Describe** a double circulatory system. **Compare** the circulatory systems of fish and mammals. **Interpret** data on pressure changes in arteries, veins, and capillaries. **Describe** how heart rate is linked to activity. **Understand** that heart muscle contraction is controlled by groups of cells called the pacemakers, and what their function is. **Recognise** that artificial pacemakers are now commonly used to control heart beat. **Recognise** that techniques such as ECG and echocardiograms investigate heart action. **Recall** that heart rate can be increased by the hormone adrenaline.	**Describe** the contribution of Galen and William Harvey towards the understanding of blood circulation. **Explain** why a single circulatory system links to a two-chambered heart. **Explain** why a double circulatory system links to a four-chambered heart. **Explain** that the blood is under a higher pressure in a double circulatory system than in a single circulatory system. **Describe** the cardiac cycle and interpret associated graphs and charts. **Explain** the sequence of contraction of the atria and ventricles, and semilunar and atrio-ventricular valves. **Describe** how the pacemaker cells coordinate heart muscle contraction **Interpret** data from ECG and echocardiograms..
B5c **Recognise** that there are many heart conditions and diseases. **Describe** reasons for blood donation. **Recall** that there are different blood groups. **Describe** the function of blood clots and appreciate that they sometimes occur abnormally. **Recall** that anti-coagulant drugs can be used to reduce clotting.	**Explain** the consequences of a hole in the heart, damaged heart valves and a blocked coronary artery. **Recognise** that there are heart assist devices as well as heart transplants. **Describe** the processes of blood donation and transfusion. **Recall** haemophilia as an inherited condition in which the blood does not easily clot. **Recall** that drugs such as warfarin, heparin, and aspirin are used to control clotting. **Describe** the process of blood clotting.	**Explain** how a hole in the heart results in less oxygen in the blood. **Understand** why an unborn baby has a hole in the heart and why it closes soon after birth. **Explain** the advantages and disadvantages of a heart pacemaker and heart valves (over a heart transplant). **Recall** that unsuccessful blood transfusions cause agglutination (blood clumping). **Explain** how the presence of agglutinins in red blood cells and blood serum determines whether a blood transfusion is successful. **Describe** which blood groups (A, B, AB, O) have which agglutinins. **Explain** which blood groups can be used to donate blood to which other blood groups.

To aim for a grade G–E	To aim for a grade D–C	To aim for a grade B–A*	
Understand why most living things need oxygen to release energy from food. **Understand** that small simple organisms take in oxygen through their external surfaces. **Recognise** that larger, more complex animals have special organs for exchange of gases. **Understand** that a large surface area improves exchange of gases. **Describe** the functions of the main parts of the human respiratory system. **Explain** the terms breathing, respiration, inspiration (inhalation), and expiration (exhalation). **Describe** the direction of exchange of carbon dioxide and oxygen at the lungs and in tissues. **Recall** that there are many conditions and diseases of the respiratory system.	**Recognise** that the methods of gaseous exchange of amphibians and fish restrict them to their habitats. **Understand** the process of ventilation (breathing) in humans. **Explain** the terms tidal air, vital capacity air, and residual air as part of the total lung capacity. **Explain** how gaseous exchange occurs within alveoli by diffusion between air and blood. **Describe** how the respiratory system protects itself from disease. **Recognise** that there are lung diseases with different causes and **describe** each disease. **Describe** the symptoms of asthma and its treatment.	**Explain** why the methods of gaseous exchange of amphibians and fish restrict them to their habitats. **Explain** how gaseous exchange surfaces are adapted for efficient gaseous exchange. **Interpret** data on lung capacities. **Explain** why the respiratory system is prone to diseases. **Describe** what happens during an asthma attack.	B5d
Describe the position and function of the parts of the human digestive system. **Describe** the process of physical digestion. **Understand** that in chemical digestion digestive enzymes break down large food molecules. **Recognise** that food enters the blood in the small intestine and leaves in body tissues.	**Explain** the importance of physical digestion. **Explain** how carbohydrates, proteins and fats are digested by specific enzymes in the body. **Recall** that stomach acid aids protease function. **Understand** that large insoluble molecules need to be broken down into small soluble molecules. **Describe** how small digested food molecules are absorbed into the blood plasma or lymph.	**Explain** how bile, from the gall bladder, improves fat digestion. **Explain** why the pH in the stomach is acidic, and the pH in the mouth and small intestine is alkaline. **Understand** that breakdown of starch is a two-step process. **Explain** how the small intestine is adapted for the efficient absorption of food.	B5e
Explain the difference between egestion and excretion. **Name** and locate the main organs of excretion. **Recall** that the kidneys excrete urea, water, and salt in urine. **Understand** what influences the amount and concentration of urine produced. **Recall** that carbon dioxide is removed from the body through the lungs.	**Understand** the importance of maintaining a constant concentration of water molecules in blood plasma. **Describe** the gross structure of a kidney and associated blood vessels. **Explain** how kidneys work. **Understand** that urea, produced in the liver, is removed from the blood by the kidneys. **Explain** why the amount and concentration of urine produced is affected by water intake, heat, and exercise. **Explain** why carbon dioxide must be excreted.	**Explain** how the structure of the kidney tubule is related to filtration of the blood and formation of urine. **Explain** the principle of a dialysis machine and how it works in a patient. **Explain** how the concentration of urine is controlled by the anti-diuretic hormone. **Explain** how the body responds to increased carbon dioxide levels in the blood.	B5f
Describe the function of the scrotum. **Describe** the menstrual cycle. **Understand** that fertilisation and pregnancy are not guaranteed for all couples. **Understand** causes of infertility. **Recognise** the use of fertility treatment. **Understand** foetal development checks. **Name** and **locate** human endocrine glands and the hormones produced.	**Describe** the role of hormones in the menstrual cycle. **Recall** that FSH and LH are released by the pituitary gland in the brain. **Explain** treatments for infertility. **Explain** the arguments for and against infertility treatments. **Describe** how foetal development can be checked to identify conditions. **Explain** why fetal screening raises ethical issues. **Understand** that fertility in humans can be controlled.	**Explain** how negative feedback mechanisms affect hormone production. **Evaluate** infertility treatments in terms of moral issues, risks, and benefits. **Explain** how fertility can be reduced.	B5g
Recall that growth can be measured as an increase in height or mass. **Understand** that a person's final height and mass is determined by a number of factors. **Describe** the main stages of human growth and identify them on a human growth curve. **Recall** that it is sometimes necessary to replace body parts. **Recall** that some mechanical replacements are used outside the body. **Understand** that organs can be donated by living or dead donors.	**Recall** causes of extremes in height. **Describe** how diet and exercise affect growth. **Recognise** that different parts of a fetus and a baby grow at diferent rates. **Understand** why a babies' length, mass and head size are regularly monitored. **Understand** the use of average growth charts. **Explain** increased life expectancys. **Explain** problems in supply of donor organs, and use of mechanical replacements. **Explain** why donors can be living and what makes a suitable living donor. **Describe** the criteria for a dead donor.	**Recall** that human growth hormone is produced by the pituitary gland. **Describe** possible consequences of more people living longer, on a personal and national level. **Describe** problems with transplants. **Describe** the advantages and disadvantages of a register of donors. **Interpret** data on transplants and success rates.	B5h

Revising module B6

To help you start your revision, the specification for module B6 has been summarised in the checklist below. Work your way along each row and make sure that you are happy with all the statements for your target grade.

If you are not sure of any of the statements for your target grade, make a note of them as part of your revision plan. You can then work back through the relevant parts of pages 204–235 to fill gaps in your knowledge as a start to your revision.

To aim for a grade G–E	To aim for a grade D–C	To aim for a grade B–A*
B6a **Recall** the size of a typical bacterial cell. **Identify** and label parts of a flagellate bacillus as shown by *E. coli*. **Recognise** that bacteria can be classified by their shape. **Describe** how bacteria reproduce. **Understand** that bacteria can reproduce very rapidly in suitable conditions. **Recognise** that bacteria can be grown in large fermenters. **Recall** that yeast is a fungus. **Identify** and label parts of a yeast cell. **Describe** how yeast reproduces asexually. **Understand** that viruses are not living and are smaller than bacteria and fungi.	**Describe** how the parts of bacterial cells relate to their function. **Describe** the main shapes of bacteria as spherical, rod, spiral, and curved rods. **Recall** that bacteria reproduce by a type of asexual reproduction called binary fission. **Describe** aseptic techniques for culturing bacteria on an agar plate. **Describe** how yeast growth rate can be increased, its optimum growth rate being controlled by certain factors. **Describe** the structure of viruses as a protein coat surrounding a strand of genetic material. **Understand** how viruses reproduce and attack.	**Explain** how bacteria survive because some bacteria consume organic nutrients and others can make their own. **Explain** the consequences of very rapid bacterial reproduction in terms of food spoilage and disease. **Explain** reasons for the safe handling of bacteria. **Describe** how yeast growth rate doubles for every 10 °C rise in temperature until the optimum is reached. **Explain** how a virus reproduces.
B6b **Understand** that some microorganisms are pathogens. **Describe** how pathogens can enter the body. **Relate** different types of microorganisms to the disease they can cause. **Recall** that some diseases can be a major problem following a natural disaster. **Recognise** that harmful bacteria can be controlled by antibiotics.	**Understand** how the transmission of diseases by food, water, contact, and airborne droplets can be prevented. **Describe** the stages in an infectious disease. **Explain** why natural disasters cause a rapid spread of diseases. **Describe** the pioneering work of Pasteur, Lister, and Fleming in the treatment of disease. **Describe** use of antiseptics and antibiotics. **Understand** that viruses are unaffected by antibiotics. **Recall** that bacteria can develop resistance to antibiotics.	**Interpret** data on the incidence of influenza, food poisoning, and cholera. **Explain** the importance of various procedures in the prevention of antibiotic resistance.
B6c **Recall** that some bacteria are useful in food production, silage, and composting. **Describe** fermentation. **Recall** some of the drinks produced by fermentation and their sources. **Recall** that carbon dioxide is produced during fermentation. **Recall** that some products of fermentation can be further treated to produce spirits.	**Describe** the main stages in making yoghurt. **Recall** and **use** the word equation for fermentation. **Describe** the stages in brewing beer or wine. **Describe** the process of distillation, and **understand** that the process needs licensed premises.	**Describe** the action of *Lactobacillus* bacteria in yoghurt making. **Recall** and **use** the balanced chemical equation for fermentation. **Explain** the implications to the fermentation process of yeast being able to undergo aerobic or anaerobic respiration. **Interpret** data on the breakdown of sugar by yeast in different conditions. **Describe** pasteurisation and explain why this is done in the case of bottled beers. **Understand** how fermentation is limited by the effects of increasing levels of alcohol. **Understand** that different strains of yeast can tolerate different levels of alcohol.
B6d **Explain** how plants produce biomass. **Recognise** examples of fuels from biomass. **Understand** why biogas is an important energy resource in remote parts of the world.	**Describe** different methods of transferring energy from biomass. **Evaluate** different methods of transferring energy from biomass, given data. **Describe** the advantages of using biofuels. **Recall** what biogas contains.	**Explain** why the burning of biofuels does not always cause a net increase in greenhouse gas levels. **Explain** how biofuels production results in habitat loss and extinction of species.

To aim for a grade G–E	To aim for a grade D–C	To aim for a grade B–A*	
Recall how the rotting of organic material occurs and what it produces. **Recall** that biogas can be produced on a large scale using a digester. **Explain** why methane being released from landfill sites is dangerous. **Recall** that alcohol can be made from yeast and can be used as a biofuel.	**Describe** how methane can be produced on a large scale. **Describe** the uses of biogas. **Describe** how biogas production is affected by temperature. **Recall** that a mixture of petrol and alcohol is called gasohol and is used for cars in Brazil.	**Recall** that biogas containing more than 50% methane can be burnt in a controlled way, but lower percentages can be explosive. **Understand** that biogas is a cleaner fuel. **Explain** why biogas production is affected by temperature. **Understand** why gasohol is more economically viable in countries that have ample sugar cane and small oil reserves.	B6d
Describe the main components of soil. **Describe** a typical food web in a soil. **Describe** the role of bacteria and fungi as decomposers. **Explain** why soil is important for most plants. **Recognise** that earthworms can improve soil structure and fertility.	**Describe** the difference between a sandy soil and a clay soil. **Recall** what loam soil is. **Recall** what humus is. **Describe** simple experiments to compare soils. **Interpret** data on soil food webs. **Explain** why some life in soil depends on a supply of oxygen and water. **Explain** the importance of humus in the soil. **Explain** why earthworms are important to soil structure and fertility.	**Explain** how particle size affects the air content and permeability of soils. **Explain** the results of soil experiments in terms of mineral particle size and organic matter content. **Explain** why aerating and draining will improve soils. **Explain** why neutralising acid soils and mixing up soil layers is important. **Recognise** Charles Darwin's work on the importance of earthworms.	B6e
Recognise the wide variety of microorganisms living in water. **Recognise** that plankton are microscopic plants and animals. **Recall** that phytoplankton are capable of photosynthesis and are producers in aquatic food chains and webs. **Understand** that plankton have limited movement and show seasonal variations. **Recall** what affects the variety and numbers of aquatic microorganisms. **Recognise** various pollutants of water. **Analyse** data on water pollution.	**Explain** the advantages of life in water. **Explain** the disadvantages of life in water. **Describe** how factors affecting photosynthesis vary in different conditions. **Interpret** data on seasonal fluctuations in phytoplankton and zooplankton. **Explain** how sewage and fertiliser run-off can cause eutrophication. **Describe** how certain species of organisms are used as biological indicators.	**Explain** the problems of water balance caused by osmosis. **Describe** the action of contractile vacuoles in microscopic animals such as amoeba. **Interpret** data on marine food webs. **Understand** that grazing food webs are most common in the oceans. **Explain** the accumulative long-term effect of PCBs and DDT on animals such as whales.	B6f
Describe everyday uses of enzymes. **Recall** that biological washing powders do not work at high temperature and extremes of pH. **Describe** how people with diabetes test their urine for the presence of glucose **Recall** how some enzymes can be immobilised. **Recall** that immobilised enzymes on reagent sticks can be used to measure glucose levels in blood.	**Describe** the enzymes in washing powders. **Explain** why biological washing powders work best at moderate temperatures. **Describe** how sucrose can be broken down. **Recognise** that, when sucrose is broken down by enzymes, the product is much sweeter. **Describe** how enzymes can be immobilised. **Explain** the advantages of immobilising enzymes.	**Explain** why the products of digestion will easily wash out of clothes. **Explain** why biological washing powders may not work in acidic or alkaline tap water. **Explain** how foods are sweetened using invertase. **Explain** the condition of lactose intolerance. **Explain** the principles behind the production of lactose-free milk.	B6g
Define genetic engineering. **Understand** that genes from one organism can work in another. **Describe** the process of genetic engineering. **Recall** that bacteria can be genetically engineered to produce useful human proteins, **Describe** how these bacteria can be grown in large fermenters to produce proteins. **Recall** that a person's DNA can be used to produce a DNA fingerprint. **Understand** that this can identify a person.	**Recall** that the new type of organism produced by genetic engineering is called a transgenic organism. **Describe** the main stages in genetic engineering. **Recall** that the cutting and inserting of DNA is achieved using enzymes. **Describe** how bacteria can be used in genetic engineering to produce human insulin. **Interpret** data on DNA 'fingerprinting' for identification. **Describe** the arguments for and against the storage of DNA fingerprints.	**Explain** why genes from one organism can work in another. **Explain** how restriction enzymes work. **Recall** that bacteria have loops of DNA called plasmids in their cytoplasm. **Explain** how plasmids can be used as vectors in genetic engineering. **Recall** that assaying techniques are used to check that a new gene has been transferred. **Describe** the stages in the production of a DNA fingerprint.	B6h

Glossary

accommodation Change in the shape of the lens of the eye to focus on near or distant objects.

active immunity Immunity that comes from making your own antibodies.

active transport Process that can move substances across cell membranes from low concentrations to high concentrations (against the concentration gradient). Active transport uses energy and is carried out by carrier proteins in the cell membrane.

adaptation Feature of an organism's body that helps it to survive in its environment.

addiction The body is dependent on a drug and will not function properly without it.

addictive Describes a drug that makes you need to keep taking more of the drug in order for your body to work properly.

adolescence 11–15 years of age.

aerobic Using/in the presence of oxygen.

afferent arteriole Blood vessel that carries blood into the glomerulus.

agglutinins Antigens (proteins) on the surface of red blood cells, that determine your blood group.

allele Version of a gene.

amniocentesis Test that may be carried out on pregnant women to test for fetal abnormality (problems with the fetus).

anaerobic Withough using/not in the presence of oxygen.

antibiotic Chemicals, usually made by fungi or bacteria, that can be used as medicines to kill other fungi or bacteria in an infected person or animal.

antibody Special protein in the body that can destroy a particular pathogen.

antigen Special protein on the surface of a cell such as a pathogen. The body makes antibodies to an antigen.

antiseptic Solution that kills microbes.

artery Blood vessel with a thick muscular wall that carries blood under high pressure from the heart to the organs.

aseptic technique Technique used for microbiological work and for tissue culture. It ensures everything is clean and sterilised so no unwanted microorganisms grow.

asexual reproduction Reproduction without gametes/sex cells, using mitosis.

atria Top chambers of the heart that receive blood from veins.

auxin Plant hormone that causes shoots to bend.

AVN Atrioventricular node, part of the pacemaker in the heart that transmits electrical impulses to the central part of the heart.

bacteria Single-celled microorganism, 1–5 μm in length. The DNA is not enclosed in a nucleus. Bacterial cells have cytoplasm, a cell membrane, and a cell wall.

balanced diet Diet that has the right amount of proteins, fats, carbohydrates, vitamins, minerals, water, and fibre and gives you enough energy.

base pairs Pairs of DNA bases; A pairs with T and C pairs with G.

base triplet Sequence of three DNA bases in a gene, that specifies a particular amino acid's position in the protein.

battery farming Farming technique in which large numbers of animals are reared indoors.

benign Describes a tumour that does not spread through the body.

binocular vision Seeing with two eyes focussing on an object at the same time.

biodiversity Measure of how many different types of organism live in in area.

biofuel Fuel such as wood or ethanol, derived from biological materials that absorb carbon dioxide while they are growing, so their use is less harmful to the environment than burning fossil fuels.

biological control Using a natural predator to control a pest, instead of chemical pesticides.

biological washing powders Soap powders containing enzymes.

biomass Dry mass of an organism. Also plant matter that can be used as a fuel.

biotechnology An industry developing ways of using microorganisms for industrial processes.

blood vessels Tube-like structures that carry blood throughout the body. Blood vessels include arteries, veins, and capillaries.

bone Living tissue that makes up most of the skeleton; bone is rigid.

breathing Movements of the rib cage and diaphragm that cause air to enter and leave the lungs.

brewing Production of alcoholic drinks by the process of fermentation.

bypass surgery Operation in which a piece of blood vessel is taken from elsewhere in the body and transplanted to bypass blocked coronary arteries, supplying the heart muscle with blood.

cancer Body disorder where cells keep on dividing and form a tumour.

capillary Small blood vessel with a very thin wall and narrow diameter. Capillaries allow exchange of substances between cells and blood.

carbohydrase Enzyme that digests carbohydrates to simple sugars.

carbon cycle Process by which carbon moves between the living and non-living world in a cyclic flow.

carbon sink Something that stores carbon, such as the oceans.

carcinogen Substance that causes cancer.

carrier protein Protein, such as haemoglobin, that carries something.

cartilage Living tissue that occurs in the skeleton; cartilage is tough and elastic.

catalyst A substance that speeds up a reaction without being used up in the reaction.

central nervous system The brain and spinal cord.

childhood 2–11 years of age.

chlorophyll Green substance found in chloroplasts, where light energy is trapped for photosynthesis.

chromosome Structure in a cell nucleus that consists of one molecule of DNA that has condensed and coiled into a linear structure.

classification Sorting organisms into groups according to their characteristics.

clone Organism that is a direct genetic copy of another organism.

community All the populations of organisms that live together and interact in the same area.

competition The struggle between organisms to get sufficient resources to survive.

complementary base pairing Pairing between DNA bases; A with T and C with G. Their shapes fit together; they are complementary.

concentration gradient Difference in concentration of a substance from one region to another.

conception Formation of a fertilised egg; the start of pregnancy.

conservation Methods of protecting habitats and species.

constrict Get narrower.

consumer Organism that consumes food made by other organisms – an animal.

contamination Process by which microbes infect a host.

contraception Preventing conception (preventing pregnancy).

cross breeding Mating between two genetically different organisms within the same species.

decay Breakdown of organic matter, such as dead organisms or food, by microbes.

decomposers Organisms that break down dead material.

deficiency disease Disease caused by not eating enough of a particular nutrient.

deficiency symptom Unhealthy symptoms shown by plants that lack essential minerals, or by animals that lack essential minerals or vitamins in their diet.

denatured State of a protein when its shape has altered and it can no longer carry out its function.

depressant Drug that depresses the activity of the nervous system.

diastolic pressure Blood pressure in the arteries when the heart relaxes.

diffusion The spreading of the particles of a gas or a substance in solution, resulting in a net movement from a region of high concentration to a region of lower concentration. The bigger the difference in concentration, the faster the diffusion happens.

diploid Describes a cell that has a nucleus with two sets of chromosomes; a body cell.

disease Condition caused by part of the body not functioning properly.

distillation Process by which alcohol from the fermentation process is made more concentrated by heating and cooling.

distribution Detail of where species are found over the total area where they occur. For example, woodlice may have a high distribution under a log.

DNA fingerprint Analysis of parts of the DNA of an individual for comparison with that of other individuals/DNA samples to find out whether someone committed a crime, or to establish whether individuals are related.

DNA ligase Enzyme used to repair DNA strands in genetic engineering, incorporating a new gene.

dominant Describes a characteristic that is expressed even if only one allele for it is present.

donor Someone who donates (gives) something, such as an organ in an organ transplant.

drug Chemical that alters the way your brain or body works.

ECG Electrocardiogram, a trace measured by doctors that shows the electrical activity in the heart.

ecosystem System including the organisms in an area and how they interact, along with non-living conditions such as rainfall, temperature, and the soil.

effector Organ such as a gland or muscle that responds to a stimulus.

efferent arteriole Blood vessel that carries blood away from the glomerulus.

efficiency Carrying out a process, such as producing food, with the minimum loss of energy.

egestion The removal of waste that has passed through the body and not been digested.

emphysema Disease of the lungs that causes breathlessness and lack of energy.

endangered species Species whose numbers are dangerously low.

energy The ability to do work in the body to maintain life.

enzyme Biological catalyst made of protein; enzymes catalyse chemical reactions in living organisms.

enzyme technology Use of enzymes as catalysts in industry.

epidemic Sudden outbreak of a disease that affects many people in a country.

eutrophication Excessive growth of plants and algae in water contaminated with nitrates.

evaporation Changing of a liquid into a vapour (gas).

evolution Gradual change in an organism over time.

excretion The removal of waste materials produced by the reactions of the body.

exponential growth In a population, exponential growth means that the more individuals there are in a population, the faster it grows.

extinction Occurs when a species is unable to survive, and all its members die.

fermentation Process by which a substance is broken down chemically through the action of yeast or bacteria. Used in production of yoghurt and alcoholic drinks.

fertilisation The joining of the male and female gametes to make a new individual.

fertiliser Chemical that promotes plant growth when added to the soil.

fibrin Insoluble protein that forms a blood clot when you cut yourself.

fish farm Technique in which fish are bred and reared in large cages in rivers or the sea.

focus Forming a clear image on the retina.

food chain A way of showing what organisms eat, showing the flow of food and energy from one organism to the next.

food web Several interlinked food chains.

fungi Organisms with cells containing a membrane, cytoplasm, a nucleus, and a cell wall. The fungal cell wall is made of chitin rather than the cellulose of a plant cell wall.

gamete Sex cell. Male gametes are sperm. Female gametes are eggs. Gametes have half the normal number of chromosomes.

gene Part of a chromosome that has the code for controlling a particular characteristic.

gene Length of DNA that codes for a characteristic/protein.

gene mutation Change in the sequence of bases in a gene. A mutation may result in the gene coding for a different protein/ characteristic.

gene therapy Inserting a copy of a functioning (allele of a) gene into a person's cells to treat a genetic disorder. It does not permanently alter the genotype and the inserted genes do not pass to offspring.

generalists Organisms that are adapted to a range of habitats.

genetic code Coded information within the DNA of an organism that controls the proteins made by its cells, so controlling its characteristics and how it develops.

genetic engineering Also called recombinant DNA technology. Permanently changing the genetic make-up of an organism by inserting gene(s) into its DNA.

genetically modified Organism that has had its genetic makeup altered using genetic engineering.

genitals Primary sexual characteristics – the external features that show the sex of an individual.

genotype The genetic makeup of an individual.

geotropic Describes the response of plants to the direction of the pull of gravity. Roots grow towards the pull of gravity and shoots grow away from it.

gland Structure that makes a useful substance for the living organism.

glasshouse Structure where plants can be grown under controlled conditions.

glomerulus Knot of capillaries in the nephron of the kidney, where filtration occurs.

growth Increase in size, usually with increase in cell numbers.

growth curve Graph showing the growth of an organism or population of organisms over time.

gullet Another name for oesophagus; in the digestive system, a tube that leads from the mouth to the stomach.

habitat Place where an organism lives.

haemoglobin Soluble protein that also contains an iron atom. Found in red blood cells. It carries oxygen from lungs to respiring tissues.

haploid Describes a cell that has a nucleus with only one set of chromosomes; a sex cell.

heart Muscular organ that pumps blood around the body.

heterozygous Describes an individual with different alleles for a gene/characteristic.

homeostasis Keeping the body's internal environment in a steady state.

homozygous Describes an individual with both alleles the same for a particular gene/ characteristic.

hormone Chemical made by a gland and carried in the blood to its target organ(s).

host An organism that has been invaded by a virus or parasite.

humus Component of soil formed by decomposition of dead organic material.

hybrid Organism resulting from two parents of different species.

hydroponics Growing plants in greenhouses without soil. Their roots are sprayed with a solution that contains the minerals they need.

immobilised Describes enzymes that are attached to an inert substance to make them more stable and easier to use.

immune system Your body's system that fights infections, invoving white blood cells and antibodies.

immunisation Medical procedure to make people immune to a particular disease. It involves introducing dead or inactivated pathogens into the body to stimulate the immune system to make antibodies.

immunity You have immunity when you have made antibodies to a pathogen, so that next time the body can make them again quickly, and the pathogen does not make you ill.

immunosuppressant drugs Drugs that suppress the immune system, given to recipients of a transplanted organ to reduce the risk of rejection.

incidence The number of new cases of a disease in a population in a period of time.

indicator species Species that survive best at a certain level of pollution, and give an idea of the level of pollution.

infancy First 2 years of life.

invertebrate Animal without a backbone.

ionising radiation Short wavelength radiation in the electromagnetic spectrum, such as X-rays and gamma rays.

IVF In vitro fertilisation – method to help achieve pregnancy in women who cannot become pregnant naturally.

key Series of questions used to classify organisms.

kingdom Major subdivision in the classification of living organisms, eg the plant kingdom.

lactic acid Chemical made from the incomplete breakdown of glucose during anaerobic respiration.

large intestine Part of the digestive system where water and some minerals are absorbed into the blood.

leaf Plant organ specialised for photosynthesis.

legume Plant that has nitrogen-fixing bacteria in its roots.

life expectancy Number of years people might normally be expected to live.

ligament Tissue that holds bones together at a joint.

limiting factor Factor such as carbon dioxide level, light, or temperature, that will affect the rate of photosynthesis if it is in short supply. Increasing the limiting fator will increase the rate of photosynthesis.

lipase Enzyme that digests fats to fatty acids and glycerol.

long sight Being able to see distant objects clearly, but not near objects.

lumen Space in a blood vessel through which blood flows.

lung capacity Total volume of the lungs, made up of vital capacity plus residual air.

macrophage White blood cell that ingests foreign particles and pathogens.

malignant Describes a tumour that spreads through the body.

maturity Having reached full adult height; usually occurs at around 18–20 years for males and 16 years for females.

meiosis Type of cell division that occurs to form sex cells, resulting in four genetically different cells.

meristems Areas within plants where the cells can divide.

microbes Microorganisms such as bacteria and yeast.

micron Unit of measurement representing a few thousandths of a millimetre.

microorganism Small organism visible under a microscope. Bacteria and viruses are microorganisms.

minerals Substances such as nitrates, phosphates, potassium, and magnesium compounds taken up by plants from the soil, that are needed for plant growth.

miscarriage Early ending of pregnancy, when the fetus dies.

mitosis Type of cell division that occurs in body cells, resulting in two genetically identical cells.

monocular vision Seeing with each eye independently, with eyes on the sides of the head.

monohybrid cross Cross between individuals in which the inheritance of one characteristic is being studied.

mouth First part of the digestive system, where food enters; contains salivary glands.

mRNA Messenger RNA: a single-stranded molecule that carries a copy of the genetic code from the cell nucleus to the ribosomes in the cytoplasm, where proteins are assembled.

muscle fatigue Inability to carry out any metabolic processes. Fatigued muscle cannot contract.

mutation Change in the structure of a gene

mutualism A relationship in which both organisms benefit from each other.

natural selection Mechanism by which species evolve. The best adapted individuals are more likely to survive and reproduce, and so pass on their genes.

needles Reduced leaves with a small surface area.

neurone Cell that carries electrical impulses, sometimes called nerve cell.

niche An organism's way of life.

nitrates Compounds of nitrogen and oxygen that are absorbed by plants from the soil.

nitrogen cycle Process by which nitrogen moves between the living and non-living world in a cyclic flow.

nuclear transfer cloning Technique used to clone an organism from genetic material taken from a differentiated adult cell of another organism. Dolly the sheep was made by nuclear transfer cloning.

oesophagus Another name for gullet; in the digestive system, a tube that leads from the mouth to the stomach.

old age Over about 60–65 years of age.

optimum Best.

organ Collection of different tissues working together to perform a function within an organism; examples include the stomach in an animal and the leaf in a plant.

organic farming Farming method that does not use intensive techniques and minimises the use of artificial chemicals.

osmosis Movement of water across a partially permeable membrane from an area of high water concentration (a dilute solution) to an area of low water concentration (a concentrated solution).

ovulation Release of a mature egg from an ovary.

oxygen debt Lack of oxygen in muscle cells. Oxygen is needed to oxidise lactic acid in the muscle to carbon dioxide and water.

oxyhaemoglobin Haemoglobin with oxygen atoms attached.

palisade layer Tissue made up of tall columnar cells containing chloroplasts, where the majority of photosynthesis occurs in a leaf.

pandemic Epidemic that sweeps across continents or across the whole world.

parasite Organism that lives in another organism (the host) and gains nutrients and shelter from the host. Pathogens are parasites.

parasitism A relationship in which one organism benefits at the expense of another.

partially permeable membrane Membrane that has small pores through which small molecules such as water can pass, but not larger molecules such as proteins.

passive immunity Immunity that comes from having antibodies injected into your body.

pathogen Organism that can cause infectious disease.

peripheral nervous system Nerves carrying information from sense organs in your body to the central nervous system, and from the central nervous system to effectors.

permeable Allows substances to pass through it freely.

pesticide Chemical used in agriculture to kill pests of crops or livestock, such as caterpillars, fleas, or weeds. Some pesticides are specific – insecticides kill insects and herbicides kill weeds.

phenotype The observable characteristics of an organism.

phloem Plant tissue made up of living cells that has the function of transporting food substances through the plant.

photosynthesis Process by which plants build carbohydrates from carbon dioxide and water, using sunlight energy.

phototropic Describes the response of plants to the direction of light. Shoots grow towards light and roots grow away from light.

pituitary gland Gland in the brain that secretes hormones, including antidiuretic hormone (ADH).

plankton Microscopic organism living in water.

plasma Fluid that makes up blood, along with blood cells. Plasma is mainly water with dissolved chemicals such as enzymes, glucose, amino acids, hormones, and wastes.

plasmid Loop of bacterial DNA used in the process of genetic engineering.

platelets Small structures in the blood that are involved in blood clotting.

pollutant Substance released into the environment that can cause harm to living things, including humans.

pollution Contamination of the environment by harmful substances.

population The number of organisms of a species in a given area.

population The number of organisms of a species in a given area.

pore Small opening. Pores on the surface of a leaf allow water and gases to move in and out of the leaf.

potometer Apparatus used to measure the rate of transpiration by measuring the uptake of water by a plant.

predator Animal that hunts and kills other animals for food.

preservation Processes that prevent food or other organic material from decaying, such as canning, freezing, drying, or salting.

prey Animal that is hunted by predators.

producer Organism that produces food – a plant.

protease Enzyme that digests proteins to fatty amino acids.

protein synthesis Building up proteins from long chains of amino acids, in the order determined by the sequence of bases in DNA.

proteins Large molecules (polymers) made of many amino acids joined together. Proteins have many functions, including structural (as in muscle), hormones, antibodies, and enzymes.

pulse Expansion and recoil of the arteries as blood is pumped into them from the heart. The pulse gives a measure of the heart rate.

pyramid of biomass A way of showing the biomass of organisms at each link in a food chain.

pyramid of numbers A way of showing the numbers of organisms at each link in a food chain.

random By chance/not ordered.

rate of photosynthesis How quickly a plant is photosynthesising. The rate is affected by factors including carbon dioxide levels, light, and temperature.

rate of transpiration How quickly a plant is losing water by transpiration (evaporation from its leaves). The rate is affected by factors including humidity, air movement, and temperature.

receptor Cell or sense organ that detects stimuli.

recessive Describes a characteristic that is only expressed when two alleles for it are present.

recycling Movement of elements from one place to another, being used in different ways time after time, such as carbon passing between living and non-living components and being reused.

red-green colour blindness Inherited condition in which people cannot tell the difference between red and green.

reflex action Fast automatic response of the body to a potentially dangerous stimulus, coordinated by the spinal cord.

rehabilitation Process of getting the body to function normally again without a drug that you are addicted to.

rejection Attack on a transplanted organ by the immune system within a recipient's body.

relationship Interaction between organisms of different species that live together and affect each other.

residual air Air that remains in the lungs after you have breathed out fully.

resistance Ability of a microorganism to withstand the effects of antibiotics and not be killed by them.

resource Something that an organism needs to survive, such as food, space, or oxygen.

respiration Process by which living things release energy from carbohydrates, also producing water and carbon dioxide.

response Action of the body brought about by a stimulus.

restriction enzymes Enzymes used to make very specific cuts through DNA in the process of genetic engineering.

ribosomes Structures in the cytoplasm of a cell, where proteins are assembled.

sampling Counting a small number of a large total population in order to study its distribution.

SAN Sinoatrial node, part of the pacemaker in the heart that produces electrical impulses which cause the atria to contract.

selective breeding Breeding programme that uses artificial selection, in which organisms with desired characteristics are chosen and interbred to produce offspring with even more desirable characteristics.

selective reabsorption Absorption of useful substances such as glucose from the kidney tubule back into the blood.

sex hormones Hormones that control sexual features; oestrogen and progesterone in females, and testosterone in males.

sexual reproduction Reproduction involving the joining of gametes from two (usually unrelated) parents.

short sight Being able to see near objects clearly, but not distant objects.

small intestine Part of the digestive system where chemical digestion is completed and absorption occurs.

specialists Organisms that are well adapted to a particular type of habitat.

species Group of organisms that are similar and capable of interbreeding.

specific Enzymes are specific; they act on only one substrate.

spines Reduced leaves with a small surface area and a pointed end.

spontaneous Happens of its own accord, with no outside influence.

stem cell Undifferentiated cell that can divide by mitosis and is capable of differentiating into any of the cell types found in that organism.

sterile Free of microbes.

sticky end Cut end of a DNA strand produced by the action of a restriction enzyme, which makes a staggered cut through a DNA strand leaving a few unpaired bases exposed.

stimulus Change in the environment, such as a temperature change, that you respond to.

stomach Part of the digestive system where physical and chemical digestion occur.

stomata Pores on the surface of a leaf that allow water, carbon dioxide, and oxygen to move in and out of the leaf.

structural protein Protein that makes up the structure of an organism; for example, collagen is a structural protein that occurs in skin, bone, and blood vessel walls.

substrate Substance acted upon by an enzyme in a chemical reaction. The substrate molecules are changed into product molecules.

surface area The area of the surface of something, such as a leaf. A small surface area reduces water loss.

sustainable Describes the use of resources for/by humans without harming the environment.

sustainable development Using resouces to meet human needs without harming the environment.

symptom How you feel when you have a disease, eg headache, feeling sick. Do not confuse symptoms with signs – signs are the measurable changes when you have a disease, such as increased temperature or a rash.

synapse Small gap between one neurone and another, or between a neurone and a muscle cell.

synovial joint Joint in the body that is lubricated by synovial fluid, such as the knee, hip, and elbow. The joint also has a synovial membrane, ligaments, and cartilage.

systolic pressure Blood pressure in the arteries when the heart contracts.

target organ Organ or part of the body that responds to a particular hormone.

tendon Tissue that connects muscles to bones; tendons are tough and do not stretch.

thrombosis Blood clot in a blood vessel.

tidal volume Volume of air breathed in during one breath.

tissue Group of cells of similar structure and function working together, such as muscle tissue in an animal and xylem in a plant.

tissue culture Method used to grow lots of genetically identical plants from the meristem tissue of one plant.

tissue match Matching the tissues of donor and recipient in a transplant operation, so that their antigens are similar and there is less chance of rejection.

tolerant You become tolerant to a drug when your body no longer responds to it to the same degree, and you need more of the drug for the same response.

toxin Poison.

transgenic Describes bacteria that have taken up an engineered plasmid (carrying a new gene) in the process of genetic engineering.

translocation Transport of sugars made in photosynthesis in the leaf to areas of the plant that store them or use them.

transmission Process by which microbes are spread.

transpiration Movement of water up through the xylem of a plant, from the roots up to the leaves.

transpiration stream Continuous flow of water up through the xylem of a plant, from the roots up to the leaves where it evaporates.

traumatic Physically damaging.

trophic level Feeding level in a food chain.

tropism Growth response to a stimulus. The direction of the stimulus determines the direction of the growth response.

tumour Mass of cells.

urea Substance produced in the liver from the breakdown of amino acids, and removed by the kidneys in the urine.

valve Structure that allows one-way movement of a fluid and prevents backflow.

variation Differences between individuals.

vasoconstriction Narrowing of blood vessels to regulate body temperature.

vasodilation Widening of blood vessels to regulate body temperature.

vector Carrier, eg of a disease-causing organism.

vein Blood vessel with a thin wall and large lumen that carries blood at low pressure from the organs back to the heart.

ventilation Movement of air into and out of the lungs, brought about by inhaling and exhaling.

ventricles Lower chambers of the heart that contract and force blood out of the heart into the arteries.

vertebrate Animal with a backbone.

viruses Extremely small infectious agents that can only be seen with an electron microscope and can only live or reproduce inside a host cell.

vital capacity Maximum volume of air you can breathe in, plus your tidal volume, plus the maximum volume of air you can breathe out after taking a big breath in.

withdrawal symptoms If you are addicted to (dependent on) a drug, when you stop taking it you get unpleasant symptoms such as pain and tremors.

xylem Plant tissue made up of dead cells that has the function of transporting water and dissolved substances through the plant.

yield Amount of product or useful substance.

zonation A gradual change in the distribution of species across a habitat.

zygote (Usually) diploid cell resulting from the fusion of an egg and a sperm.

Index

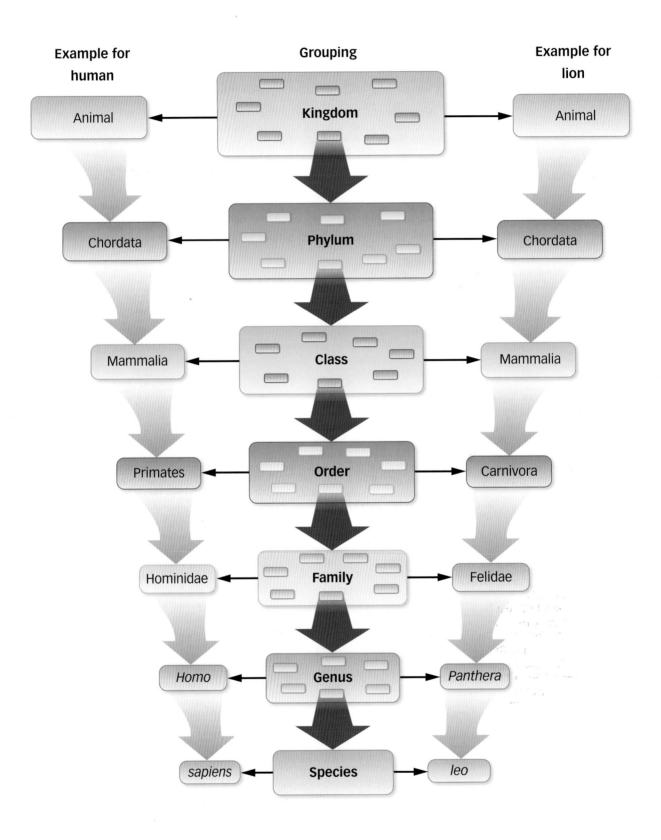

Acknowledgements

The publisher and authors would like to thank the following for their permission to reproduce photographs and other copyright material:

p8 Martyn F. Chillmaid/SPL; p9 Martyn F. Chillmaid/SPL; p10T Laurence Gough/Istockphoto; p10B Gordon Scammell/Alamy; p11 Chris Pearsall/Alamy; p13 www.frankhuelsboemer.de; p14 Mark Turnball/SPL; p15 Biophoto Associates/SPL; p16 Elena Schweitzer/Shutterstock; p17 Mauro Fermariello/SPL; p19R Sheila Terry/SPL; p19L Sheila Terry/SPL; p20M Medical RF.com/SPL; p20R Steve Gschmeissner/SPL; p20L Sinclair Stammers/SPL; p22M SPL; p22R Dr. Gladden Willis, Visuals Unlimited/SPL; p22L Thierry Berrod, Mona Lisa Production/SPL; p24M Medical RF.com/SPL; p24R Medical RF.com/SPL; p24L Corbis/Photolibrary; p25 Art Wolfe/SPL; p26 Medical RF.com/SPL; p27 Iofoto/Shutterstock; p28 Bob Gibbons/SPL; p30T Dan Sams/SPL; p30B Photostock-Israel/SPL; p31 Maximilian Stock Ltd/SPL; p32 Tony McConnell/SPL; p36 Jerome Wexler/SPL; p37L Dan Suzio/SPL; p37M John Kaprielian/SPL; p37R Peter Anderson/Dorling Kindersley/Getty Images; p38B Dr Gopal Murti/SPL; p38TL Mark Burnett/SPL; p38TR Mark Burnett/SPL; p39T Patrick Landmann/SPL; p39B Stuart Wilson/SPL; p40 Picsfive/Fotolia; p41 Eye of Science/SPL; p42 SPL; p44L Annabella Bluesky/SPL; p44R Eye of Science/SPL; p51 Power and Syred/SPL; p52L Dr Keith Wheeler/SPL; p54TL John Beatty/SPL; p54TR Mark Phillips/SPL; p54BL Alexis Rosenfeld/SPL; p54BR Alexis Rosenfeld/SPL; p55 Todd Lammers/Istockphoto; p57R Georgette Douwma/SPL; p57L Matthew Oldfield/SPL; p61 Curt Maas/AGStockUSA/SPL; p63 Lynn Mclaren/SPL; p65 Pasieka/SPL; p68L Peter Chadwick/SPL; p68R Mark Newman/SPL; p69T Jeff Lepore/SPL; p69B Peter Chadwick/SPL; p73 T-Service/SPL; p75L Tony Camacho/SPL; p75R K Jayaram/SPL; p76TL National Library of Medicine/SPL; p76TR Science Source/SPL; p76B SPL; p77 Michael W. Tweedie/SPL; p79 Brian Bell/SPL; p81TR Ken Harris; p81L Simon Harris/Photolibrary; p81MTR Massimo Brega/The Lighthouse/SPL; p81MBR Dr J. Bloemen/SPL; p81BR Duncan Shaw/SPL; p89 Michael Webb, Visuals Unlimited/SPL; p90R Dr Gopal Murti/SPL; p90L Adrian T. Sumner/SPL; p91 Medical RF.com/SPL; p92 Patrick Landmann/SPL; p94BL John Bavosi/SPL; p94TL BSIP, Laurent/Louise Eve/SPL; p94TR Thomas Deerinck, NCMIR/SPL; p94BR AJ Photo/SPL; p95 Tim Vernon/SPL; p98 Pascal Goetgheluck/SPL; p99L Lawrence Migdale/SPL; p99R Peter Menzel/SPL; p100T Tony Craddock/SPL; p100B Joe McDonald, Visuals Unlimited/SPL; p105 Herve Conge, ISM/SPL; p107 Eye of Science/SPL; p108 National Cancer Institute/SPL; p109 Ed Reschke, Peter Arnold Inc./SPL; p110L Medical RF.com/SPL; p110R Dr. Gladden Willis, Visuals Unlimited/SPL; p114 GustoImages/SPL; p115 Sinclair Stammers/SPL; p116T Golden Rice; p116B Subbotina Anna/Shutterstock; p118 Chris Knapton/SPL; p119 Robert Brook/SPL; p120 Mark Thomas/SPL; p127 Hank Morgan/SPL; p128TL Martyn F. Chillmaid/SPL; p128TR Martyn F. Chillmaid/SPL; p129B Philippe Psaila/SPL; p131T Chris Dawe/SPL; p131B Doug Sokell, Visuals Unlimited/SPL; p134 Scott Sinklier/AGStockUSA/SPL; p136R Veronique Leplat/SPL; p136L Power and Syred/SPL; p140 Biophoto Associates/SPL; p141BR Michael Abbey/SPL; p141TR Michael Abbey/SPL; p141L Steve Gschmeissner/SPL; p142 Gavin Kingcome/SPL; p143R Power and Syred/SPL; p143L Eye of Science/SPL; p145T Dr Jeremy Burgess/SPL; p145BL Dr Jeremy Burgess/SPL; p145BR Dr Jeremy Burgess/SPL; p146L Martyn F. Chillmaid/SPL; p146R Cindy Hughes/Shutterstock; p150B Robert Brook/SPL; p150T GustoImages/SPL; p151L Sinclair Stammers/SPL; p151R Bob Gibbons/SPL; p152 Astrid & Hanns-Frieder Michler/SPL; p153T1 Courtesy of Crown Copyright Fera/SPL; p153T2 Peter Menzel/SPL; p153T3 Volker Steger/SPL; p153T4 BSIP, Marigaux/SPL; p153T5 David Munns/SPL; p153T6 David Munns/SPL; p153T7 Maximilian Stock Ltd/SPL; p154 Debra Ferguson/AGStockUSA/SPL; p156T Simon Fraser/SPL; p156B Rosenfeld Images Ltd/SPL; p157T Simon Fraser/SPL; p157B Peter Menzel/SPL; p158 Mauro Fermariello/SPL; p159 Dr Jeremy Burgess/SPL; p165 Tek Image/SPL; p166 Michael Patrick O'Neill/SPL; p169 Living Art Enterprises, Llc/SPL; p173T Gary Carlson/SPL; p173B D. Varty, Ism/SPL; p174T GustoImages/SPL; p174B BSIP, Raguet H/SPL; p176 SPL; p178 David Aubrey/SPL; p179 Claude Nuridsany & Marie Perennou/SPL; p182 Steve Gschmeissner/SPL; p183T Moredun Animal Health Ltd/SPL; p183B Medical RF.com/SPL; p185 Eye of Science/SPL; p191 Medical RF.com/SPL; p192T AJ Photo/SPL; p192B CC Studio/SPL; p195 Jacob Wackerhausen/SPL; p196R Life In View/SPL; p196L Hank Morgan/SPL; p203 Patrick Landmann/SPL; p204T Eye of Science/SPL; p204MT Scimat/SPL; p204MB London School of Hygiene & Tropical Medi-Cine/SPL; p204B SPL; p206T Science VU, Visuals Unlimited/SPL; p206B David M. Phillips/SPL; p207L Pasieka/SPL; p207R Cavallini James/SPL; p208 Mark Clarke/SPL; p209T Tim Vernon, Lth Nhs Trust/SPL; p209BL Adam Hart-Davis/SPL; p209M Dr H.C. Robinson/SPL; p209BR Adam Hart-Davis/SPL; p210 Lowell Georgia/SPL; p212TL Custom Medical Stock Photo/SPL; p212TR Science Source/SPL; p212B Chris Ware/Stringer/Hulton Archive/Getty Images; p214L BSIP Chassenet/SPL; p214R Trevor Clifford Photography/SPL; p215T Robert Brook/SPL; p215BL Rosenfeld Images Ltd/SPL; p215BR Mark Sykes/SPL; p217 Mike Bentley/Istockphoto; p224 Julie Dermansky/SPL; p225 BSIP Martin Pl./SPL; p226MB Saturn Stills/SPL; p226T Rosenfeld Images Ltd/SPL; p226MT Veronique Leplat/SPL; p226M BSIP Chassenet/SPL; p226B Heike Brauer/Istockphoto; p227 Jonathan Hordle/Rex Features; p228 Saturn Stills/SPL; p229 Jim Amos/SPL; p231R Maximilian Stock Ltd/SPL; p231L ISM/Photolibrary; p234T Pasieka/SPL; p234B Dra Schwartz/Istockphoto; p235 Martin Shields/SPL.

Cover image courtesy of TED KINSMAN/SCIENCE PHOTO LIBRARY.

Illustrations by Wearset Ltd, HL Studios.

The publisher and authors are grateful for permission to reprint the following copyright material:

p19 Survival rates for cancer for men, Cancer Research UK, http://info.cancerresearchuk.org/cancerstats/survival/fiveyear/, October 2010.

Although we have made every effort to trace and contact all copyright holders before publication this has not been possible in all cases. If notified, the publisher will rectify any errors or omissions at the earliest opportunity.

OXFORD

UNIVERSITY PRESS

Great Clarendon Street, Oxford OX2 6DP

Oxford University Press is a department of the University of Oxford.
It furthers the University's objective of excellence in research,
scholarship, and education by publishing worldwide in

Oxford   New York

Auckland   Cape Town   Dar es Salaam   Hong Kong   Karachi
Kuala Lumpur   Madrid   Melbourne   Mexico City   Nairobi
New Delhi   Shanghai   Taipei   Toronto

With offices in
Argentina   Austria   Brazil   Chile   Czech Republic   France   Greece
Guatemala   Hungary   Italy   Japan   Poland   Portugal   Singapore
South Korea   Switzerland   Thailand   Turkey   Ukraine   Vietnam

Oxford is a registered trade mark of Oxford University Press
in the UK and in certain other countries.

British Library Cataloguing in Publication Data

Data available

ISBN 978-0-19-913568-4

10 9 8 7 6 5 4

Printed in China by Printplus

Paper used in the production of this book is a natural, recyclable product
made from wood grown in sustainable forests. The manufacturing process
conforms to the environmental regulations of the country of origin.